Disaster Deferred

DISASTER DEFERRED

How new science is changing our view of earthquake hazards in the Midwest

Seth Stein

Columbia University Press
New York

Columbia University Press

Publishers Since 1893

New York Chichester, West Sussex

Library of Congress Cataloging-in-Publication Data

Stein, Seth.

Disaster deferred : how new science is changing our view of earthquake hazards in the Midwest / Seth Stein.

p. cm.

Includes bibliographical references and index.

ISBN 978-0-231-15138-2 (cloth : alk. paper)—ISBN 978-0-231-52241-0 (ebook)

1. Earthquake prediction—Technological innovations. 2. Earthquake prediction—Middle West. 3. Earthquake Hazard analysis—Middle West. I. Title.

QE538.8.S74 2011

551.220978—dc22 2010029738

Columbia University Press books are printed on permanent and durable acid-free paper.

This book is printed on paper with recycled content.

Printed in the United States of America

c 10 9 8 7 6 5 4 3 2 1

For my daughter, Rachel, who told me that
"Science is today's way of keeping magic in the world."

Science invites us to let the facts in, even when they don't conform to our preconceptions. It counsels us to carry alternative hypotheses in our heads and see which best fit the facts. It urges on us a delicate balance between no-holds barred openness to new ideas, however heretical, and the most rigorous skeptical scrutiny of everything— new ideas and established wisdom.

—Carl Sagan, *The Demon-Haunted World: Science as a Candle in the Dark*

CONTENTS

Disaster Deferred

Chapter 1

Threshold

The prize is the pleasure of finding the thing out.

—Physicist Richard Feynman, 1918–1988

Stretched out on a sun-warmed rock, I admired the hawks circling lazily in the bright-blue sky. It was a perfect October day in 1997 in Petit Jean State Park, high in Arkansas's Ozark Mountains. About 50 yards away, I could see a five-foot yellow-and-orange wooden tripod topped by a shiny metal disk that looked like a large Frisbee. The disk was an antenna receiving radio signals from Global Positioning System (GPS) satellites orbiting thousands of miles above the earth (fig. 1.1). An expensive, high-precision GPS receiver about the size of a personal computer recorded the signals and used them to find the antenna's latitude and longitude to incredible accuracy.

In the early morning chill, I'd carefully set up the tripod over a metal marker drilled into solid rock. Doing this involved sighting through a lens to position the tripod over the marker and adjusting the tripod legs to make sure the antenna was level. This complicated sequence felt like an intelligence test that I'd slowly and only barely passed.

The GPS receiver was doing a very simple thing—measuring its location—using incredibly complicated space technology. Fellow geologists and I had installed 24 markers like this one over a large area in the central U.S. We'd measured their positions in 1991 and 1993 and were now doing it again.

Our goal was to learn more about the mysterious zone of earthquakes called the New Madrid seismic zone. It's named after the town of New Madrid, pronounced "MAD-red," in the area of southeastern Missouri known as the "Bootheel." The zone includes parts of Missouri, Arkansas, Tennessee, Kentucky, Illinois, and Indiana. In 1811 and 1812, large earthquakes here shook the central U.S., and small earthquakes continue in the zone today. A map of

FIGURE 1.1 GPS antenna at Petit Jean State Park.

the recent small earthquakes shows some major patches, which we think are mostly aftershocks of the past large earthquakes, surrounded by a diffuse "cloud" (fig. 1.2).

These earthquakes are interesting because they're in a strange place. Most big earthquakes happen at the boundaries between the great rock plates that slide around on the earth's surface. For example, the San Andreas fault in California is part of the boundary between the Pacific and North American plates. In contrast, the New Madrid seismic zone is a less active earthquake zone in the middle of the continent, within the North American plate.

Geologists know surprisingly little about what's going on here. We don't know why the earthquakes occur; when they started; if, when, and where future large earthquakes will occur; how serious a danger they pose; or how

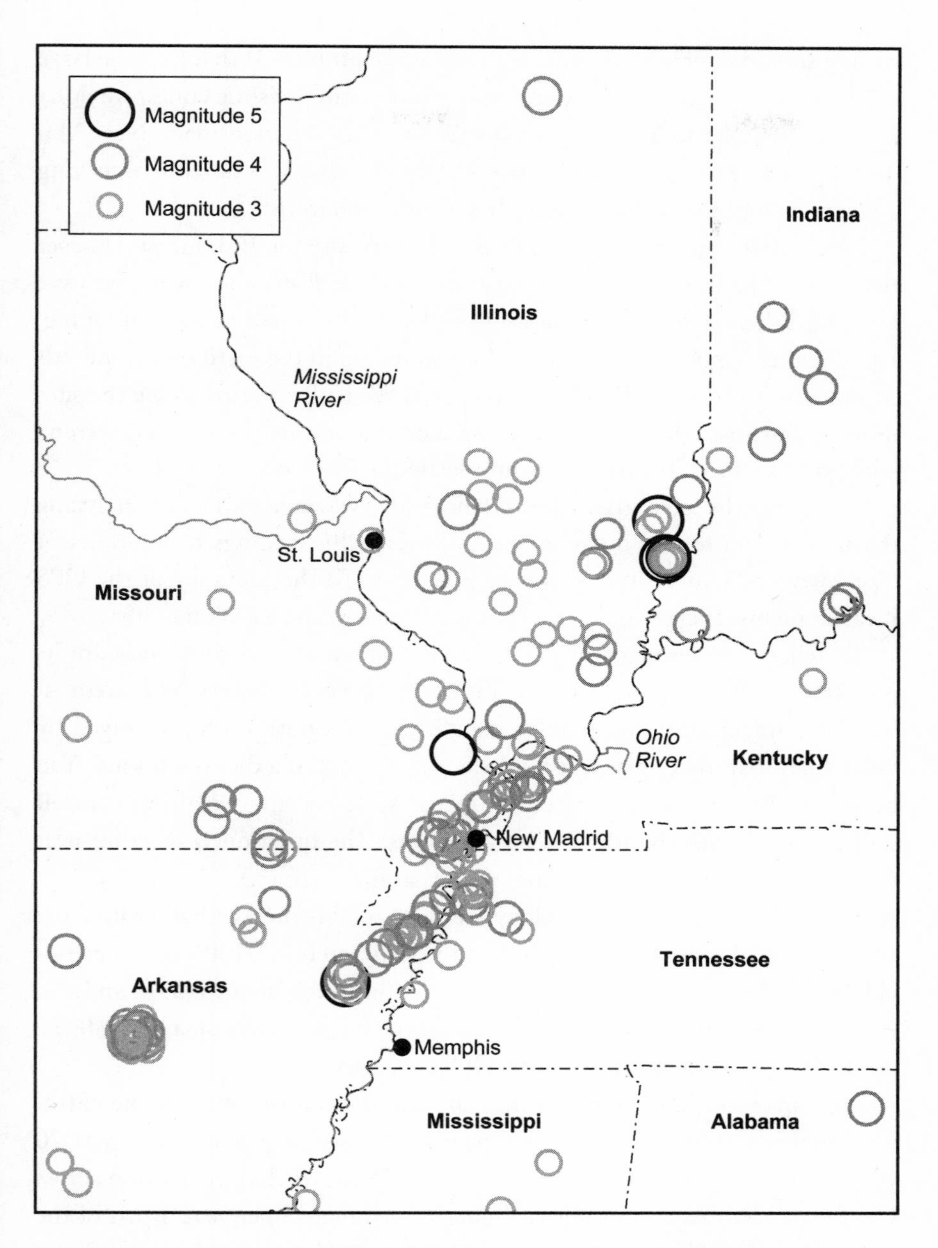

FIGURE 1.2 Locations of earthquakes between 1975 and 2008 in and around the New Madrid seismic zone. (After University of Memphis)

society should confront them. A big part of the problem is that because large earthquakes here are much rarer than in many other seismic zones, we don't yet have the data to answer these questions. This situation made New Madrid a perfect place to use the new GPS method that was quickly becoming a powerful tool for earthquake studies around the world.

Because we were recording GPS data at each site for 10 hours a day over three days, I had a lot of time in a beautiful place to think about what we were learning. Already, it looked like we were on the threshold of something big. Earthquakes happen when slow motions stored up in the earth over hundreds or thousands of years are suddenly released. We had expected to see the sites moving. Surprisingly, we weren't seeing that, but our first two surveys weren't enough to be sure. This survey would settle the question.

A few months later, graduate student Andy Newman, who was analyzing the survey data for his doctoral thesis, brought his findings to my office at Northwestern University. The result was clear. To the accuracy of the GPS measurements, the ground across the earthquake zone wasn't moving.

To tell if a monument in the ground is moving, geologists measure its position at different times and see if it changes. Because every measurement has some uncertainty, we look to see whether the position has changed by more than that uncertainty. It's like the way you tell if a diet is working. You know that there's some uncertainty in the scale because weighing yourself several times gives slightly different answers. The question is whether over time your weight changes by more than that uncertainty.

The GPS systems used in geology are so incredibly precise that we measure the motions of markers in the ground in millimeters—1/1000 of a meter—per year. Because a meter is 39.37 inches, a millimeter is about 1/25 of an inch. As figure 1.3 shows, that's about the size of the bigger letters on a dime. If the ground moves that much in a year, GPS can detect it.

This precision lets geologists measure the slow movements of the earth. That makes GPS the most important new tool we've gotten in the past 20 years for earthquake studies. For example, GPS shows that the motion across California's San Andreas fault is about 36 millimeters per year. Most of the time, the fault is "locked" by the friction between the rocks on either side, so the motion is stored up in the rock. Eventually, the stored motion overcomes the friction, and the fault moves in a big earthquake. This happens about every hundred years, so in seconds the fault moves about 3,600 millimeters, or about 12 feet!

FIGURE 1.3 GPS measurements can tell if a point on the earth moves by more than a millimeter—about 1/25 of an inch—in a year. That's the height of a capital letter on a dime.

Our GPS data for New Madrid didn't show any motion. Specifically, they showed that the ground was moving less than 2 millimeters per year. That's at least 18 times more slowly than the San Andreas. We were also pretty sure that the number would get much smaller if we kept measuring for a longer time. Already it was a lot slower than we'd expect if a big earthquake were coming any time soon. I thought of the joke in which a tourist asks a Maine farmer "Does this road go to Bangor?" and is told "Nope, stays right where it is."

During the next few months, all of us in the project talked at length about what the lack of motion might mean. Conventional wisdom was that the New Madrid area faced a major earthquake risk. The U.S. government claimed that this risk was as high as in California and was pressuring communities to make expensive preparations.

The GPS data showed that these common ideas about the New Madrid seismic zone needed serious rethinking. All of us in the project were excited. Although as scientists we're trying to solve the earth's mysteries, much of our effort actually goes into day-to-day chores like trying to fix computer programs. Most of the time we do routine studies and find answers that aren't too surprising. The most exciting times are when a project gives an unexpected answer that leads to new insight. Sometimes that's due to clever planning, but often, as in this case, it's just unplanned dumb luck. We have only a few of these moments in our careers, so we treasure them.

Because a scientist's most exciting moment of discovery is often his first, I was especially pleased for Andy. During many years of advising students, I'd learned that those who make major discoveries while in school generally go on to make others later. I think that's because they learn to spot things that don't fit into the accepted picture. Instead of forcing new data to fit into their preconceptions, they learn to think outside of conventional wisdom. We all try to do this but usually fail. Hence, most of our contributions come from the few "aha" moments when we break free. It feels like a door has just opened, and there's so much new to explore.

To working scientists like me, real science is very different from the ideal "scientific method" taught in elementary school. That ideal scientist is like a lone explorer who examines the possible paths to a clearly visible mountain, chooses the best, and presses on. Real scientists are like a mob of hikers trying to find the way to an unseen lake through dense woods full of swamps, mosquitoes, and poison ivy. We argue about which routes look best, try different ones, follow them when they seem to be working, and try others when they aren't. It's exciting and fun but also confusing and frustrating. Eventually, mostly through luck, we reach the lake, often by different routes that get there about the same time. Once we're at the lake, we argue about whether it's the right lake.

The moral is that while searching for the lake, we were all confused and going in the wrong directions about half the time. We finally got there as a group by combining many people's efforts. It's hard to say who contributed what because we're all sure that we played a key role. It's also not that important, because after relaxing in satisfaction for a while, we realize that there's a bigger lake somewhere higher up on the mountain, and it's time to get to work looking for it.

Because scientists are human, science is a very human endeavor. Scientists choose problems to study and methods to study them that reflect their interests, skills, and sense of where their efforts will yield useful new knowledge. Typically, others are exploring different aspects of similar or related problems, using different methods, and sometimes finding different results and drawing different inferences. Even starting with the same set of observations, how individual scientists interpret them depends in large part on their preconceptions. Although we'd like to be totally objective, we can't be. It's like watching a sports event when it's unclear what happened in a tricky play; fans see the result that's good for their team. Sometimes instant replay convincingly settles the question, and sometimes it doesn't.

Eventually, as scientific knowledge increases, a clear result emerges that combines many people's work over many years. Until then, scientists grappling with a problem often have differing views. There's spirited debate about the meaning of our incomplete results. This debate is crucial for progress, as described by the ancient Jewish sages' adage "the rivalry of scholars increases wisdom."

This messy process has been going on since the large earthquakes in 1811 and 1812 brought the New Madrid seismic zone to scientists' attention. At the time, geology had just begun as a science, and seismology, the branch of geology that studies earthquakes, didn't exist. As seismology evolved and knowledge about the New Madrid earthquakes increased, it became clear how unusual they are and how challenging it is to understand them. Hence, many researchers have been exploring scientific and policy issues for the area.

Many of the results I'll be talking about come from work that graduate students, friends—mostly from other midwestern universities—and I have done in the past 20 years. This isn't a formal, structured project. Geologists in general are individualists who like messy problems that don't have simple, clean solutions. Most of us view ourselves—correctly or not—as generalists with broad ranges of skills rather than as narrow specialists. Thus, instead of working in large organizations with clear hierarchies and assigned tasks, we typically work as loose groups of friends with overlapping interests. We're very informal, so faculty and graduate students work pretty much as equals. We share ideas, but all involved have their own take on what's going on.

The results we'll discuss also show how empirical geology is. The earth is too complicated to have fundamental laws that predict what's going on, the way physics does. Instead, geologists observe what's happening and try to make sense of it. What we find often surprises us and forces us to change our ideas. The earth regularly teaches us humility in the face of the complexities of nature. Thus, although we take our science seriously, we generally don't take ourselves too seriously.

This geologists' outlook makes it easier to communicate our science to the public. Many of us do a lot of education and outreach and enjoy it. Whether through class and public lectures, the media, or just talking to people, I find wide interest in how our planet works and how its workings affect humanity.

This book grew out of that interest. It's an overview of studies of the science and hazards of the New Madrid earthquakes. These studies involve many people, each of whom views the subject differently. The book is written from my perspective, developed through 35 years of studying earthquakes around the world and 20 years of thinking about New Madrid.

I'll show how scientists study the questions surrounding the New Madrid earthquakes, how far we've gotten, and what we still have to do. I'll explain what we know and what we don't, what we suspect, and which is which. Discussing these questions involves going into some concepts that can seem a little complicated, because the earth is complicated.

Understanding these concepts lets us look at issues that scientists, engineers, policy makers, and the public are struggling with. These issues are interesting scientifically and have practical significance because billions of dollars are involved. Moreover, if you live in the Midwest, understanding what we're learning will reduce your fear of earthquakes.

We've learned a lot about the New Madrid earthquakes in the past 20 years, so the picture coming out is very different from older ideas. Still, there's a lot we don't yet know. That's typical in studying the earth. A college student can take years of classes in many other sciences before encountering topics that the instructor admits aren't understood. In the earth sciences, the first courses present many unsolved fundamental questions about our planet. An earth science instructor is like the apocryphal medical school dean who tells incoming students: "Half of what we will teach you in the next four years is wrong. The problem is that we don't know which half."

I hope readers—especially students considering possible careers—take away a sense of the excitement, fun, and opportunities in science. Earth science, in particular, offers the challenge of working on interesting and important problems like earthquakes in the middle of continents, where what we learn will help us live better with a complicated and active planet.

Chapter 2

The Day the Earth Stood Still

This is the silliest thing I've ever been to.

—TV cameraman

Media Circus

December 3, 1990, was a very strange day in the little town of New Madrid, Missouri. Dozens of television trucks filled the streets (fig. 2.1). Reporters and camera crews roamed around, filming and interviewing residents, tourists, and each other. Residents filmed the television crews. T-shirt and barbecue sales were brisk. Hap's Bar and Grill did a booming business that proprietor Jack Hailey described as an all-day party. Dick Phillips, mayor of the town of about 3,200 people, described the day as a three-ring media circus.

Usually, New Madrid is a small, quiet community in Missouri's southeast corner, far from St Louis's big-city world. It has a southern flavor, with friendly people, a mostly warm climate, and fertile soil well suited for agriculture.

Tourists driving along the Mississippi River on Interstate 55 sometimes stop in New Madrid to watch the barge traffic along one of the world's great trade waterways. An observation deck looks over a major river bend, the control of which was crucial during the Civil War. In 1862, Union naval and land forces won a major victory a few miles upriver at Island No. 10 (the tenth island south of the point where the Ohio and Mississippi rivers meet). The battle opened the way for General Ulysses Grant's advance downriver to Vicksburg that put the Mississippi under Union control and split the Confederacy, leading President Abraham Lincoln to proclaim "The Father of Waters again goes unvexed to the sea."

FIGURE 2.1 Television trucks near Main Street in New Madrid before the predicted earthquake. (AP/Wide World Photos)

Since the war, most outside interest in New Madrid has come from geologists who are trying to learn more about the big earthquakes that occurred nearby in 1811 and 1812. Residents of the area sometimes feel small earthquakes but regard them as just part of life.

Everything changed in December 1989. Iben Browning, an eccentric business consultant, announced that there was a 50% chance of a major earthquake with a magnitude of about 7 in the New Madrid seismic zone within a few days of December 3, 1990.

Earthquake predictions, like predictions of other disasters, are nothing new. They've probably been around as long as people have felt earthquakes, but they surface disproportionately after big earthquakes capture the public's interest. In this case, Browning's prediction included the claim that he had predicted the magnitude 6.9 Loma Prieta earthquake that had struck the San Francisco Bay area on October 17, 1989, two months earlier. That earthquake drew national attention because it disrupted a nationally televised World Series baseball game and caused 63 deaths and $6 billion in property damage.

Usually, disaster predictions have no effect, but this one caught on. Journalists at the main offices of most newspapers recognized the prediction as junk science and initially ignored it. However, the story was picked up by the Associated Press and slowly spread. Some news stories accepted Browning's claim to have predicted the Loma Prieta earthquake. The *St. Louis Post-Dispatch*, for example, wrote that "Mr. Browning has a good record of accurate predictions" and the *New York Times* reported that Browning "is known to have predicted the 1989 San Francisco earthquake."

By the fall of 1990, Browning's prediction had drawn considerable publicity. With varying degrees of enthusiasm, many seismologists tried to explain why the convincing-sounding prediction was nonsense. We're used to this and have the script down pat. Members of the media ask when the next major earthquake will occur in a particular place, and we explain that we can't predict earthquakes, don't know if it will ever be possible to do so, and suspect that it won't be.

As we'll discuss in Chapter 4, this failure to reliably predict earthquakes isn't from lack of effort. Geologists have been trying since seismology became a science in the late nineteenth century. Every so often, a method is proposed that in hindsight might have predicted a major earthquake. However, when the method is applied to predict future earthquakes, it yields either false positives (predictions without earthquakes) and/or false negatives (earthquakes without predictions). Watching years of failures led Hiroo Kanamori, who developed many of the methods used in modern earthquake studies, to the observation that it is very difficult to predict earthquakes before they happen.

Still, predictions keep coming. Charles Richter, who developed the magnitude scale used to measure the size of earthquakes, could have been describing Browning when he wrote of predictors "What ails them is exaggerated ego plus imperfect or ineffective education, so that they have not absorbed one of the fundamental rules of science—self-criticism. Their wish for attention distorts their perception of facts, and sometimes leads them on into actual lying."

His point was that predictors aren't skeptical enough of their predictions. As a result, seismologists have to be. When someone claims to have predicted an earthquake, the test is whether the prediction was more successful than random guessing based on a general knowledge of where earthquakes occur. After all, guessing will be right some of the time, purely by chance. It's like comparing the performance of a mutual fund to an index of the stock market as a whole. In any year, some funds beat the index, either by skill or by chance. Although the managers of these funds boast of their accomplishment, it's likely to be just luck unless they do it year after year.

Using this approach, a committee of seismologists examined Browning's prediction and found that it made no sense. Browning claimed that the earthquake would result from the tides. The gravitational attraction between the earth, sun, and moon causes tides—tiny ripples—in the solid earth, just as in the ocean. The sizes of these ripples vary, just as ocean tides are largest at new and full moon, when the sun, earth, and moon line up. Browning argued that around December 3, the sun, earth, and moon would line up, and the moon would be especially close to the earth, causing unusually large tidal forces.

The committee found that although Browning said that his method had predicted other earthquakes, he had never said where they would be. Instead, he claimed success whenever an earthquake occurred on a date for which he predicted tidal highs. For that matter, Browning didn't say why he was predicting an earthquake at New Madrid rather than anywhere else. Also, similar tidal highs in the past hadn't caused New Madrid earthquakes. In summary, Browning hadn't done better than guessing. As committee member Duncan Agnew pointed out, anyone could do as well by throwing darts at a calendar.

In addition, the committee debunked Browning's claim to have predicted the Loma Prieta earthquake. The "prediction" turned out to have been that around that date magnitude 6 earthquakes would occur somewhere in the world. This didn't prove much because, on average, an earthquake of this size occurs somewhere on earth every three days.

The committee's report settled the matter for scientists. As we'll see in Chapter 14, there's no certain way to estimate the probability of a large earthquake. There are different ways that give different answers. The simplest is to use geological averages: If large earthquakes like those in 1811 and 1812 occur on average once every 500 years, the chance of one on any particular day is 1 in 365 × 500, or 1 in 182,500. This works out to about 0.0005%—about 100,000 times lower than the probability that Browning claimed for December 3.

Calm analysis didn't dampen the media's enthusiasm. Newspapers and TV newscasts played up the prediction and generally ignored or downplayed its failings. Few pointed out that Browning's predictions (available on tape for $100) included one that said tidal forces would cause a depression and the collapse of the U.S. government in 1992 or that Browning also blamed tidal forces and volcanic eruptions for the American Revolution, the Civil War, the social unrest that began in the U.S. in 1961, and the collapse of communism in Eastern Europe.

It's not clear why the story wasn't given the critical analysis applied to other news stories. I've heard that some newspapers' science reporters said to ignore the story, but local news desks continued to play it regardless. Only a few journalists, including Karen Brandon of the *Kansas City Star* and Mary Beth Sammons of *Crain's Chicago Business*, critically looked into the prediction and pointed out that it was nonsense.

It didn't help that November is one of the months when Nielsen ratings determine how much stations can charge local advertisers for commercials. During this time, known as "sweeps," competition for viewers is ferocious. As a result, local television coverage became almost hysterical.

For a while, I just watched the coverage with a mixture of amusement and frustration. However, my Northwestern colleagues and I got involved after watching Chicago media outlets play up the prediction. Some of the silliest coverage was by Bill Kurtis and Elizabeth Vargas on CBS's flagship station, WBBM-TV. When we tried to persuade the station to be more sensible, its representatives said that they could and would do as they pleased. This attitude didn't surprise a journalism professor I talked to who explained, "Management tends to view local news as a commercial commodity little different from entertainment." We had more luck with the newspapers, where staffers enjoy needling their broadcast competitors. After hearing our side, the *Chicago Sun-Times*'s radio and TV columnist, Robert Feder, described the WBBM stories as "an over-hyped but typically flimsy sweeps series."

By late November, much of the New Madrid area was preparing for a large earthquake. Many schools held earthquake drills, and authorities made preparations like parking fire trucks outside. Businesses closed or took special precautions. Many people left the area or stocked up on chemical toilets, flashlights, generators, and other emergency supplies. In Missouri alone, homeowners spent $22 million on earthquake insurance. Insurance sold briskly even in places like Chicago that were so far from the fault that homeowners wouldn't collect even if the earthquake occurred.

As things got wackier, scientists continued trying to calm the media but increasingly realized it was hopeless. It was like watching a herd of stampeding buffalo that would only stop after falling over a cliff. We watched in resignation because most of us had nothing at stake.

An exception was Doug Wiens, a talented young seismologist at Washington University in St. Louis. Doug, whose thesis I'd advised when he was a Ph.D. student, was undergoing the review universities conduct to decide whether to grant tenure to or dismiss a junior faculty member. Trying to debunk Browning, Doug became an articulate media spokesman for seismologists. I hoped he wouldn't be in trouble in the extremely unlikely event that—by sheer dumb luck—Browning proved right.

Things were much tougher for people who had to deal with public fears. Mike Lynch, who coordinated earthquake preparedness for the state of Kentucky, held public meetings that many fearful people attended. Lynch recalls his telephone ringing almost continuously as December 3 approached, with some callers crying as they asked how far away they should move to be safe. Taking the optimistic view, a cartoon in the *San Francisco Chronicle* suggested that New Madrid would become the "next California," with nouvelle cuisine and stress reduction classes.

Eventually, December 3 arrived. Missouri Governor John Ashcroft visited New Madrid. Many schools and mines in the area closed, and reporters and television trucks descended on the town like locusts. Tom's Grill offered quake burgers. A preacher's van labeled "Earthquake or Rapture—Jesus is coming soon" blared religious messages. The members of the media were somewhat sheepish because they knew they were participating in a non-event largely of their own making. One TV news producer explained, "No one is going down there to cover an earthquake. The only story is one of fear and overreaction—which we helped create." A cameraman admitted, "This is the silliest thing I've ever been to." My favorite summary came from a young resident who told a reporter, "I think all of you are stupid for wasting your time coming down here."

ABC news invited me to join the festivities in New Madrid. I probably should have, just for fun, but opted not to. Instead, I gambled against the nightmare scenario of the earthquake happening on December 3 purely by chance and taped a segment for CNN in advance. I then went off to the American Geophysical Union's annual conference in San Francisco and watched myself on television. In the story, CNN announced, "The only shaking over the New Madrid fault is scientists shaking heads over the hype and hoopla from the big earthquake prediction. However, life in the region has been disrupted by a media stampede." James Warren of the *Chicago Tribune* added, "This is a classic example of rather miserable shoddy, imprecise reporting, which stems from this guy's somewhat goofy claim to have predicted the San Francisco earthquake."

A day later, after no earthquake happened, the story mercifully ran out of steam. Newspapers joked that the rumble you heard was the crumpling of insurance policies. The newly bought earthquake insurance policies were cancelled, and the media admitted that they had been taken in. The *Riverfront Times*, a St. Louis alternative newspaper, wrote, "The media and pols were a big easy for the quacks and maniacs. But was it as good for you as it was for them? And more important: do you still respect them in the morning? Hardly."

The Morning After

As usual when a predicted disaster doesn't happen, the media just moved on. For a few days, seismologists enjoyed "we told you so" satisfaction. Still, in the months that followed, a lot of us pondered the question: What created the climate for this silly prediction to be so readily accepted and have such dramatic impact?

Our first instinct was to blame someone else. The easy, satisfying answer was that the public, press, and authorities in the area have little experience with earthquakes. Unlike California, where large earthquakes are common, damaging earthquakes are very rare in the heartland. Hence, a prediction that would probably have been dismissed in California was taken seriously. Sociologist John Farley conducted telephone surveys in the New Madrid seismic zone before December 3 to study what he called a "pseudodisaster," a situation in which the public incorrectly believes that a disaster is either underway or imminent. Approximately 20% of those surveyed considered an earthquake around December 3 "very likely" and about 40% thought it "somewhat likely." Only about 15% regarded it as "very unlikely." In St. Louis, about 150 miles from the predicted earthquake, 25% reported planning to change their schedules near December 3. In Sikeston, Missouri, about 20 miles away from New Madrid, the number was 52%! 5% of those surveyed in St. Louis and 20% of those surveyed in Sikeston said they planned to leave the area on the dreaded day. Thus, although not everyone was taken in—some joked that the real threat was flooding on December 4 when everyone dumped the bottled water they'd stored—many residents took it seriously.

Unfortunately, on careful reflection, it became clear that the problem was closer to home. Browning's prediction was the spark that set off prepared firewood. The firewood was what federal and state agencies and even some university scientists had told the public about past earthquakes and future hazard. Much of this was wildly exaggerated to the point of being embarrassing.

The 1811–1812 earthquakes have been described in apocalyptic and mythical terms. As we'll see in later chapters, much of this is hype. The earthquakes were claimed to have been the largest ever in the contiguous U.S., which is far from the case. Many accounts say that the earthquakes rang church bells in Boston, which didn't happen. The Shawnee leader Tecumseh is said to have predicted the earthquakes though historians doubt it.

Claims of future risk are equally inflated. Even today, the U.S. Geological Survey (USGS) argues that the earthquake hazard in the New Madrid seismic zone is similar to or greater than that in California. The Federal Emergency Management Agency (FEMA) continues to issue dire warnings of impending doom and to pressure states in the area to spend billions of dollars to raise the earthquake resistance of buildings to levels similar to California's.

Like most seismologists, I'd known that a lot of this was exaggerated but never thought much about it until I started doing the GPS project. As we started realizing that the ground wasn't moving the way we'd expect if a big earthquake were on the way, I began to wonder. In 35 years, I've studied many earthquake zones on earth and even some on the moon. Compared with any of those I'd studied or knew of, New Madrid seemed to have the highest ratio of hype to what we really know. How did this come about? How could intelligent people have talked themselves into believing that the earthquake danger in the Midwest was so huge?

Chapter 3

Think or Panic?

As soon as I hear "everybody knows" I start asking
"Does everybody know this, and how do they know it?"

—Seismologist David Jackson

A teacher's most satisfying moments are when students "get it." One of my favorites was when a student said, "I heard on TV that last weekend four people were killed in accidents caused by the ice storm. Then I thought about what we did in data analysis class and wondered whether that was more than in a typical weekend."

That's exactly what scientific education is supposed to accomplish. Although courses teach facts about faults, earthquakes, waves, rocks, minerals, and so on, they're only useful once students learn to think critically. Thus, our goal as educators is also to show students how to ask the right questions: What are the data? How good are they? What's the interpretation? Does it make sense? Is there an alternative?

Critical thinking is basically thinking for oneself. It's a scientist's most powerful tool. Without it, even modern equipment and complicated computer programs can lead us down the wrong paths. I think that's what happened with the New Madrid earthquakes.

Earthquake Hazards and Risks

Critical thinking is hard for anyone under the best circumstances. It's especially hard for scientists dealing with earthquakes. We're either reacting to the immediate aftermath of one or thinking about hypothetical effects of a future one.

Part of most seismologists' experience—often beginning while we're students—is dealing with the media and public after earthquakes. There's a brief communication window before interest turns elsewhere. We're famous for 15 minutes, as Andy Warhol said, or even a bit longer. After the December 26, 2004, Indian Ocean earthquake and tsunami that killed more than 200,000 people, media interest remained high until Hollywood's dream couple, Brad Pitt and Jennifer Aniston, announced their separation on January 7. On the *Today* show, host Katie Couric explained, "I actually called our news desk on Saturday and said: 'I know that we have this tsunami going on, and all these people, but is it true that they broke up?' "

Early in their careers, seismologists often fumble these opportunities because geological education doesn't include media training. Soon, we learn what to do: look at the camera, talk in short sound bites, and avoid technical terms. We try to talk about the earthquake, usually before we really have enough information. We try to present the science about earthquakes and how they happen. We also try to explain the hazards earthquakes pose and how society can best reduce them.

Explaining earthquake hazards involves two different and seemingly opposite goals. Seismologists try both to explain earthquake hazards and give a sensible perspective on how they compare to other hazards society faces. Thinking calmly about disasters and deaths is uncomfortable but necessary. As long as seismology has been a science, we've been discussing these issues. Hence, during the 1950s and 1960s, Charles Richter was an early advocate for earthquake resistant construction in California while pointing out, "I don't know why people in California or anywhere worry so much about earthquakes. They are such a small hazard compared to things like traffic."

Probably every seismologist since has faced questions about earthquake hazards. Sometimes people call me and ask whether they shouldn't visit Disneyland or take a job in California because of the danger of earthquakes. My advice is to take the trip and buckle their seat belts.

Comparing the typical number of deaths per year in the U.S. from various causes, as shown in table 3.1, shows this point.

U.S. earthquakes have caused an average of about 20 deaths per year since 1812. The precise number depends mostly on how many died in the 1906 earthquake that destroyed much of San Francisco, which isn't known. The analysis starts after 1812 because we don't know if anyone was killed in the big New Madrid earthquakes.

TABLE 3.1 Typical Number of Deaths per Year in the U.S.

Heart attack	733,834
Cancer	544,278
Motor vehicles	43,300
AIDS	32,655
Suicide	30,862
Liver disease/Cirrhosis	25,135
Homicide	20,738
Falls	14,100
Poison (accidents)	10,400
Drowning	3,900
Fire	3,200
Bicycle accidents	695
Severe weather	514
Animals	191
In-line skating	25
Football	18
Skateboards	10

As you'd expect, these numbers show that earthquakes aren't a major cause of death in the U.S. Although earthquakes are dramatic and can cause major problems, many more deaths result from familiar causes like drowning, fires, or bicycle accidents. Severe weather disasters are about 25 times more dangerous than earthquakes. Earthquakes rank at the level of in-line skating or football.

This relatively low ranking arises because most U.S. earthquakes do little harm, and even those felt in populated areas are commonly more of a nuisance than a catastrophe. In most years, no one is killed by an earthquake (fig. 3.1). About every 40 years, an earthquake kills more than 100 people, and the 1906 San Francisco earthquake is thought to have killed about 3,000 people. This pattern arises because big earthquakes are much less common than small ones, and large numbers of deaths occur only when a rare big earthquake happens where many people live. Other natural disasters, like hurricanes, have the same effect, with rare large events doing the most damage. Because people remember dramatic events and don't think about how rare they are, it's easy to forget that more common hazards are much more dangerous.

Earthquakes can also cause major property damage. The 1994 Northridge earthquake under the Los Angeles metropolitan area caused 58 deaths and $20 billion in property damage. The latter number, though large, is about 10% of the U.S. annual loss due to automobile accidents.

The reason earthquakes are only a secondary hazard in the U.S. is that large earthquakes are relatively rare in heavily populated areas, and buildings

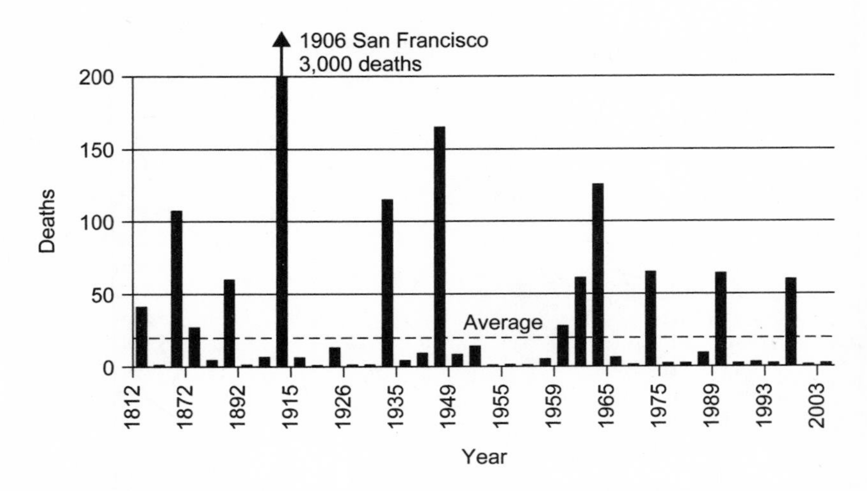

FIGURE 3.1 U.S. earthquake deaths for the years from 1812 to 2003.

in the most active areas like California are built to reduce earthquake damage. Earthquakes are a bigger problem in some other countries where many people live near plate boundaries. During the past century, earthquakes worldwide caused about 11,500 deaths per year. Although the statistics are imprecise, major earthquakes can be very destructive. Estimates are that the 1990 northern Iran earthquake killed 40,000 people and that the 1988 Spitak (Armenia) earthquake killed 25,000. Even in Japan, where modern construction practices are used to reduce earthquake damage, the 1995 Kobe earthquake caused more than 5,000 deaths and $100 billion in damages. The highest property losses occur in developed nations where more property is at risk, whereas fatalities are highest in developing nations. For example, the January 2010 Haiti earthquake killed more than 200,000 people.

Earthquakes pose an interesting challenge to society because they cause infrequent, but occasionally major, fatalities and damage. As a result, society needs to think carefully about what to do.

Most deaths from earthquakes happen when buildings collapse, so it's said that "earthquakes don't kill people; buildings kill people." Thus, the primary defense against earthquakes is designing buildings that shouldn't collapse. We can make buildings as safe as we want, but deciding how safe is tough. It involves trying to estimate how much of a hazard future earthquakes pose and choosing a level of safety that makes sense. That's because making buildings safer raises construction costs and diverts resources from other uses that could do a lot of good or save more lives. The more a community spends, the safer it will be in future earthquakes, but the more it gives up for this safety. Because the costs involved are typically in the range of 1–10% of building costs and sometimes more, the total cost can be billions of dollars over hundreds of years.

In thinking about these issues, it helps to make a pedantic-sounding but useful distinction between "hazards" and "risks." The hazard is the natural occurrence of earthquakes and the resulting ground shaking and other effects. The risk is the danger the hazard poses to life and property. The hazard is a geological fact, but we don't know how big it is and so have to make our best estimates. In contrast, the risk is affected by human actions like where people live and how they build. Areas of high hazard can have low risk because few people live there. Areas of modest hazard can have high risk due to large populations and poor construction. The important idea is that earthquake risks can be reduced by human actions, even though hazards can't.

All of this makes it hard to decide how to best prepare for earthquakes. Part of the problem is that people often don't think critically. We're often more afraid of unfamiliar but minor risks than familiar and much greater ones. As a result, we sometimes don't prepare in a thoughtful way that would do more good. For example, the U.S. spends billions of dollars on measures that might reduce the risk of rare terrorist attacks while not taking the much cheaper measures that would certainly reduce the much larger number of lives lost every year in automobile accidents.

This fear of the rare and unknown sometimes makes us act irrationally. Sometimes, as with the Iben Browning earthquake prediction, we panic over unlikely disasters. This happens when fear of a potentially real and poorly understood problem grows wildly—via hype and worst-case assumptions—into a panic. Because the common feature of these panics is a failure to think critically, looking at what happened in other cases gives some ideas about what to avoid.

Disaster Panics

Every year, influenza—the flu—kills about 30,000 Americans. People take flu for granted, and most don't bother getting a shot that should reduce the risk from that year's strain of the virus. However, exotic flu strains regularly trigger panics.

In 1976, some scientists and physicians at the U.S. government's Centers for Disease Control (CDC) warned of an upcoming swine flu epidemic. Although many of the staff thought an epidemic was unlikely, their views were ignored. In the *Washington Post*'s words "The rhetoric of risk suffered steady inflation as the topic moved from the mouths of scientists to the mouths of government officials." Soon the head of the CDC discussed the "strong possibility" of an epidemic. Staff in the U.S. Department of Health, Education, and Welfare announced that "the chances seem to be 1 in 2" and the secretary of the department told the White House that "projections are that this virus will kill one million Americans in 1976."

Some scientists, including Albert Sabin who developed the oral polio vaccine, called for stockpiling the vaccine but only using it if an epidemic occurred. These calm voices had no effect. Perhaps due to the upcoming election, President Gerald Ford's administration launched a program to vaccinate the entire population (fig. 3.2). Forty million Americans were vaccinated at a cost of millions of dollars before the program was suspended due to reactions to the vaccine. About 500 people had serious reactions, and 25 people died,

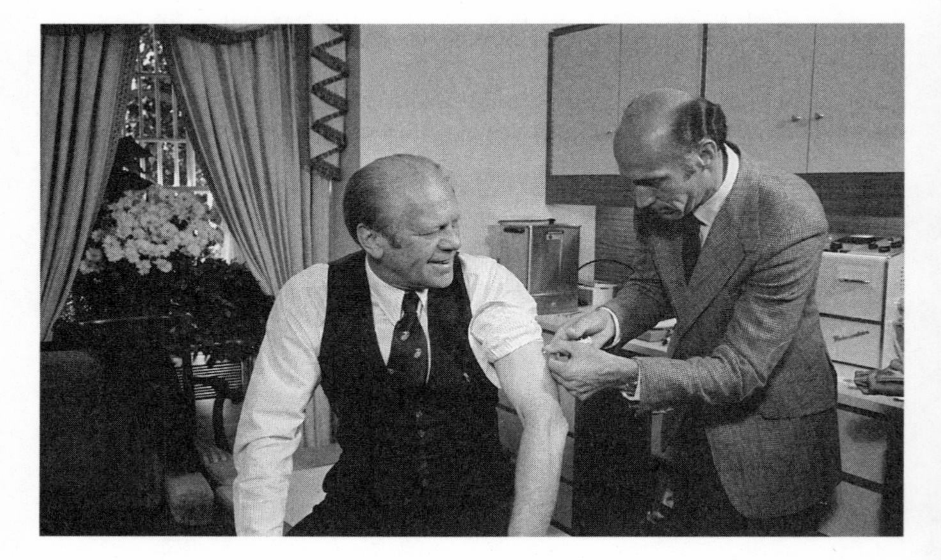

FIGURE 3.2 President Gerald Ford is vaccinated against swine flu in 1976. (Gerald R. Ford Library)

whereas only one person died from swine flu. Ford lost the election, so the only immediate winners were the companies that made the vaccine. The fact that the dreaded "aporkalypse" didn't materialize helped the public health system respond somewhat more thoughtfully to new exotic strains like the 2003 bird flu and the 2009 swine flu.

In the late 1990s, some computer experts warned of an impending catastrophe on January 1, 2000. Supposedly, computer systems would fail because they represented dates in their memory using only two digits. Hence, the year 2000, known as Y2K, would not be distinguished from 1900. Books and the media warned of upcoming disaster due to the collapse of the financial system, public utilities, and other institutions of technological society (fig. 3.3). Many warned that Y2K would cause TEOTWAWKI—"the end of the world as we know it."

The hype grew despite some calm voices, including Microsoft's Bill Gates who predicted no serious problems and blamed "those who love to tell tales of fear" for the scare. *Newsweek* carried a cover story titled, "The Day the World Shut Down: Can We Fix the 2000 Computer Bug Before It's Too Late?" TV programs, like one starring *Star Trek*'s Leonard Nimoy that is now available on YouTube, explained the impending catastrophe. Survivalists stockpiled food, water, and guns. Y2K insurance policies were written.

Senator Daniel Moynihan wrote to President Bill Clinton urging that the military take command of dealing with the problem. The U.S. government established the President's Council on Year 2000 Conversion headed by a "Y2K czar." Special legislation was passed and government agencies led by FEMA swung into action. FEMA, in turn, mobilized state emergency management agencies.

On the dreaded day, the "millennium bug" proved a bust. Nations and businesses that had prepared aggressively watched sheepishly as a few—and only minor—problems occurred for those that had made little or no preparation. However, the estimated $300 billion spent made Y2K preparations very profitable for the computer industry. Y2K programming work shipped from the U.S. fueled the growth of India's profitable outsourcing industry at the expense of American programmers' jobs. On the plus side, there was a long-term benefit in that claims of impending computer-related disasters are now treated more calmly.

Despite swine flu and Y2K, panic usually doesn't happen. We're constantly warned of dangers: sharks, SARS, West Nile virus, terrorism, computer viruses,

FIGURE 3.3 One of many books advising how to survive the predicted Y2K disaster.

Lyme disease, killer bees, communists, flesh-eating bacteria, fire ants, mad cow disease, etc. Although some people overreact, common sense generally prevails.

Even so, many researchers interested in the fear of risks think that panics are becoming more common. We're losing our perspective. Sociologist Barry Glassner, author of *The Culture of Fear*, points out that "crime rates plunge, while surveys show Americans think that crime rates are rising. At a time when people are living longer and healthier, people are worried about iffy illnesses."

Researchers feel that such fears are becoming more serious for two reasons. First, groups with agendas feed them. In *False Alarm: The Truth About the Epidemic of Fear*, Marc Siegel from the New York University School of Medicine explains, "Bureaucrats use fear by playing Chicken Little. They can claim at every turn that a disaster may be coming, thereby keeping their budgets high and avoiding blame should disaster really strike." Cass Sunstein of the University of Chicago points out in *Risk and Reason: Safety, Law, and the Environment* that an official "may make speeches and promote policies that convey deep concern about the very waste spill that he actually considers harmless." Unfortunately, bureaucrats aren't the only culprits. Siegel argues "many of our scientists who inform us do us a disservice by overdramatizing their concerns."

Second, the media play up fears. In most cases—as in the Browning prediction—reporters know the story is hype, but the dynamics of the business encourage them not to say so. There's an incentive not to ask hard questions. As a journalist explains in *False Alarm*, "I don't like reporting the overhyped stories that unnecessarily scare people, but these are my assignments." Siegel writes, "Many news teasers use the line 'Are you and your family at risk?' The answer is usually no, but that tagline generates concern in every viewer and keeps us tuned in. If we didn't fundamentally misunderstand the risk, we probably wouldn't watch."

We know that what we're told is mostly hype. That's why we're not surprised when the predicted disaster of the week turns out much smaller than predicted or doesn't happen at all. We don't hold it against the government officials who issued the warning or the media who trumpeted it. No one apologizes or issues retractions. Soon there's a warning about another disaster, and we get just as excited about it.

I think disaster stories keep coming because people enjoy being a little scared. We like disaster movies, Halloween, and roller coasters. Although real dangers like car accidents are too real to be fun to think about, it's fun to worry about small but exotic risks that we know are unlikely to actually hurt us. Many people like disaster worries so much that they hate to give them up. There's no harm—unless we lose perspective and start to believe the danger is real.

Groupthink

Many panics have common features. They involve a potentially real but unfamiliar problem, whose magnitude is hard for both specialists and the public to assess. This situation calls for critical thinking, but for some reason it doesn't happen. Instead, fear exaggerates the problem and causes massive overreaction.

The overreactions result from a combination of factors, including the government wanting to appear to be doing something, the media hyping the story, and entrepreneurs seeing opportunities to profit. A bad dynamic called "groupthink" kicks in.

Groupthink is so common that political scientists and management experts have studied it to understand how groups sometimes make bad decisions. The studies find that in groupthink, members of a group start with an absolute belief that they're doing the right thing. They focus on the unlikely worst case and don't consider the harm their actions cause. They convince one another using specious arguments, pressure others to agree, refuse to consider clear evidence that contradicts their views, and ignore outside advice. They more they talk to one another, the more convinced they become.

It's like the Middle Eastern tale about a merchant whose stall is bothered by pesky children. To get rid of them, he makes up a story that free figs and nuts are being handed out at the river. Once the boys leave, he thinks, "If there are free figs and nuts at the river, what am I doing here?" and runs off.

Some of the best studied cases of groupthink involve political decisions, including the U.S. government's decisions to launch the failed invasion of Cuba at the Bay of Pigs, go to war unsuccessfully in Vietnam, and launch the space shuttle Challenger that exploded. There are lessons to be learned from these events. One is that talented people can fall into groupthink. President Dwight Eisenhower, who approved the Bay of Pigs, had led the carefully planned and successful invasion of Europe during World War II. The Vietnam planners had seemed to be a capable bunch; David Halberstam's book about them is called *The Best and the Brightest.* A second lesson is that groups can learn from mistakes. Learning from the Bay of Pigs fiasco, the U.S. did much better in the 1963 Cuban missile crisis. President John Kennedy's advisors carefully thought through different options, looked at the possible outcomes, and came up with a successful policy.

Are Scientists Better?

Scientists instinctively smile when thinking about panics and groupthink. These happen because bureaucrats, politicians, and the media overreact. This couldn't happen to us, we think. We're trained to think critically.

Unfortunately, history shows that scientists make similar mistakes. A famous example deals with the number of chromosomes in the human body. For 30 years, researchers using microscope photos to count the chromosomes in human cells were convinced there were 48: 23 pairs plus the X and Y chromosomes that determine sex. In 1956, better analysis methods showed that there are 46. That's good, because apes have 48. For years, researchers had followed the "consensus" that everyone "knew" was right. This kind of groupthink, which takes place over years and includes people who might never have met, is called a "bandwagon effect."

How the bandwagon effect occurs is demonstrated by the history of measurements of the speed of light. Because this speed is the basis of the theory of relativity, it's one of the most frequently and carefully measured quantities in science. As far as we know, the speed hasn't changed over time. However, from 1870 to 1900, all the experiments found speeds that were too high. Then, from 1900 to 1950, the opposite happened—all the experiments found speeds that were too low! This kind of error, where results are always on one side of the real value, is called "bias." It probably happened because over time, experimenters subconsciously adjusted their results to match what they expected to find. If a result fit what they expected, they kept it. If a result didn't fit, they threw it out. They weren't being intentionally dishonest, just influenced by the conventional wisdom. The pattern only changed when someone had the courage to report what was actually measured instead of what was expected.

This human tendency to see what we expect to see, called "confirmation bias," is a major problem in science. As the saying goes, "It's not what you don't know that hurts you—it's what you know that isn't so." It's natural to plan research to confirm what's expected rather to look for what isn't.

Similarly, people tend not to see what they don't expect. A nice example is an experiment where people are asked to watch a video of a basketball game and count how many times one team gets the ball. Concentrating on the count, many viewers miss a person in a gorilla suit walking across the court.

The speed of light history also demonstrates another major problem: The researchers thought their measurements were more accurate than they really were. Scientists are taught to always consider the uncertainties in measurements or data. We know that we can't measure things exactly, so we report our best value and what we think its uncertainty is. If we've done this right, the true value—which we don't know—should be within the uncertainty. However, the speed of light measurements didn't work out that way. Over time, the results of successive measurements stayed about the same, and once researchers realized that there was a problem, the results changed by more than what had been thought to be the uncertainty. Thus, in scientific lingo, the researchers "underestimated their uncertainties."

Although uncertainties sound obscure, they're crucial to any scientific issue, including deciding what to do about possible future earthquakes in the New Madrid zone. Putting certain values into a computer program predicts that Memphis is at greater risk of earthquakes than much of California. The question is how uncertain that prediction is. As we'll see, the uncertainty is huge, so we can get almost any answer we want by changing the values we put in.

The problem is that we can't measure uncertainties. How can we measure how well or poorly we know something? Thus, scientists talk about "estimating," not "measuring," uncertainties. ("Guessing" would often be a better word, but it doesn't sound scientific.) To understand the history of measurements of the speed of light, we compare older measurements to newer, more accurate ones. This works in hindsight, but researchers in 1900 couldn't do it to figure out how good their measurements were.

To give just one more example, in September 2008 the National Weather Service predicted that Hurricane Ike would strike Miami. When instead it headed toward the Texas coast, the service warned that people on Galveston Island faced "certain death." In fact, fewer than 50 of the 40,000 who stayed on the island were killed. The predicted 100% probability of death was 800 times too high.

Years of experiences like these have taught scientists a hard lesson: Uncertainties are almost always greater than we think they are. The reason is simple. When we try to figure out uncertainties, we include all the sources of uncertainty we can think of. Unfortunately, there are always problems that we didn't think of.

Scientists have also learned that we can fall victim to groupthink. We've all accepted conventional wisdom, which gave a wrong answer, without critical thinking. Even worse, because the uncertainties were underestimated, we had confidence in the wrong answer. The wrong answer got locked in because we discounted the data that disagreed. Eventually, someone spotted what was wrong, slowly convinced others, and the whole house of cards collapsed.

Avoiding being trapped by conventional wisdom is tough. The best way is to be a young graduate student who doesn't know or care what older scientists think. The next best is to be a faculty member working with that kind of student. I've been lucky to have ones who said "that's stupid" when they didn't agree with me. The third best way is to constantly remind yourself not to give in to conventional wisdom. This need for critical thinking has been said in many ways, including Mark Twain's dictum that "sacred cows make the best hamburger."

Chapter 4

The Perfect Mess

Scientists need to learn to be excited by what we do without telling a premature story. We can alarm people unnecessarily. And then we're stuck with our story, right or wrong.

— Marc Siegel, *False Alarm: The Truth About the Epidemic of Fear*

Sebastian Junger's book *The Perfect Storm* describes how several different features of weather in the North Atlantic combined to produce an unusually powerful storm. That's a good analogy for how the earthquake hazard at New Madrid got exaggerated to the point that it was being called bigger than in California.

This "perfect mess" grew from interactions between groups of people focusing on different scientific and policy issues. Each group's natural and usually healthy tendencies somehow got taken to excess and fed on each other. The result was a case of groupthink in which a few people convinced themselves and then the public of the exaggerated hazard.

Although a lot of what was being said seemed unlikely, groupthink overwhelmed critical thinking. That's a shame because scientists are supposed to be skeptical. As astronomer Carl Sagan said, "extraordinary claims require extraordinary evidence." He was explaining that an unlikely claim, like that space aliens have kidnapped humans, requires very strong proof.

In this chapter, we'll look at the groups involved and what happened. Then in later chapters we'll take a detailed look at the arguments that produced the mess.

Universities

Seismologists are fascinated by earthquakes and naturally try to raise interest about them and get funding to study them. The challenge is to do this while maintaining a sense of proportion. For many years this worked for New Madrid studies.

Otto Nuttli of St. Louis University (SLU), who started the modern scientific studies of the New Madrid seismic zone in the 1970s, set the tone. The fact that St. Louis University led the way might seem strange. You might think first of the Midwest's "big-name" universities. In seismology, however, SLU boasts a proud tradition owing to its Jesuit roots. From the earliest days of seismology, Jesuit universities around the world set up seismometers—devices that measure the ground motion earthquakes produce—and exchanged data. The SLU program was founded by Father James Macelwane, who is easily recognized by his clerical collar in pictures of early 1930s seismologists.

Nuttli, a quiet and well-respected scientist, faced the challenge that the earthquakes in 1811 and 1812 happened before the seismometer was invented in the 1880s. Seismometers produce seismograms, which are records of the ground shaking produced by earthquakes. These are used to find the location of an earthquake and its magnitude, or size.

Without seismograms, what could be learned? Nuttli's solution was to analyze historical accounts of the large earthquakes, which he interpreted as showing that the biggest of the 1811–1812 earthquakes had a magnitude of about 7.4. This was big but smaller than the largest earthquakes on the San Andreas fault. He then did the same thing for smaller earthquakes that happened in the New Madrid zone in the years before seismograms became available. Combining these older data with studies of later earthquakes using seismograms gave the first good maps of the geometry of the seismic zone and the rate at which earthquakes happen there.

Although Nuttli advocated research to assess the hazard and "if necessary, take remedial steps," he didn't exaggerate the hazard. According to his colleague Brian Mitchell, "he received numerous phone calls and much correspondence from residents of the St. Louis area, some of whom might have an irrational fear of an impending earthquake. Otto would listen patiently and, if necessary, would attempt to allay their fears."

In addition to SLU, other universities in the Midwest and elsewhere are active in New Madrid research. Because New Madrid earthquakes big enough to learn much from are rare, significant progress is slow. Thus, most geology departments regard New Madrid as one of a large number of interesting questions around the world. Similarly, most scientists who work on New Madrid think it's interesting but just part of what they do. As a result, most researchers have followed Nuttli's example in taking a calm approach.

However, there have been exceptions. One was David Stewart, a faculty member at Southeast Missouri State University. He was also director of the Central United States Earthquake Consortium (CUSEC), a federally sponsored group of state emergency management agencies that regularly issues dire warnings of upcoming earthquake disasters. Stewart had been a faculty member at the University of North Carolina. In 1974, he predicted that an earthquake would strike Wilmington, near a nuclear power plant. Stewart asked the governor to close the plant and flew over the area with a psychic. After no earthquake happened and the university didn't give him tenure, Stewart moved to Missouri and promoted the New Madrid earthquake hazard. He defended Iben Browning's earthquake prediction and was widely quoted by the media during the months before December 3, 1990. His visibility faded after the earthquake didn't happen.

The most effective academic advocate of a high earthquake hazard was Arch Johnston, who as a newly graduated seismologist founded the Tennessee Earthquake Information Center (TEIC) at Memphis State University in 1978, with himself as director. Subsequently, TEIC was renamed the Center for Earthquake Research and Information (CERI), and Memphis State became the University of Memphis. Under Johnston's direction, CERI took a different approach to New Madrid earthquakes than other university geology programs. CERI focused primarily on New Madrid, which had the advantage of developing local knowledge and data but the disadvantage of taking less of a worldwide perspective. As we'll see, a lot can be learned about New Madrid using what's known about earthquakes elsewhere in North America and in other continents. What's going on makes more sense and looks a lot less scary.

Johnston rapidly put CERI in the public eye. Within a few years, he reported that the biggest 1811–1812 earthquake had had a magnitude of about 8.1. As we'll see in Chapter 8, this value would make it five times larger than

Nuttli found. Johnston also redid Nuttli's estimates of the rate of New Madrid earthquakes and in 1985 argued that there was a 40–63% probability of a smaller but strong (magnitude 6) earthquake occurring somewhere in the New Madrid zone by the year 2000.

All of this was what groups promoting the hazard wanted to hear. The larger magnitudes of the 1811 and 1812 earthquakes caught on. Johnston described them as the largest known in continental North America. *Hidden Fury*, a widely circulated video, went even further and described them nonsensically as "the most powerful series of earthquakes ever known on earth."

These claims should have raised red flags. Earthquakes happen because of the motion of the earth's plates, so it would be surprising for the biggest ones to be inside a plate, rather than at the boundaries between moving plates. That's not impossible, but it would be rare, because it's like the tail wagging the dog. As for the probabilities, we can calculate those in many ways and get almost any number we want.

U.S. Geological Survey

Another major player has been the U.S. Geological Survey (USGS). Although best known to the public for its maps, the USGS also conducts geological research like that in universities. The USGS was founded in 1879 to survey and map the vast areas of the western U.S., with an emphasis on helping the mining industry. With time, its responsibilities and staff grew. Today, the USGS, with more than 8,000 employees and a billion dollar budget, is bigger than all U.S. university geology departments combined.

The organizational complexity of the USGS is important for our story. USGS scientists function in various ways. Some work in structured projects that produce products like the earthquake hazard maps we'll discuss. Others work like university scientists with reasonable freedom as long as they're not too critical of official positions.

Thus, when I criticize USGS actions, products, or positions, I'm speaking of a small part of a large organization. Most USGS scientists I know do good science. This situation is like the way few, if any, university scientists are involved in the athletic scandals that get their schools in the news. (As the joke says, "How many basketball players does it take to change a light bulb? The whole team, and they all get an A for it.")

The USGS is part of the Department of Interior, which also includes the Bureau of Indian Affairs, National Park Service, Minerals Management Service, and the U.S. Fish and Wildlife Service. To thrive and grow in this complex organization, the USGS has needed both canny bureaucratic skill and expert public relations. A bureaucratic triumph that's important for our story was getting control of the government's seismology programs.

During the 1960s, several factors generated increased interest in earthquake studies. Two giant earthquakes happened, both of which were much bigger than the New Madrid earthquakes of 1811 and 1812. In 1960, the largest known earthquake, with a magnitude of 9.5, occurred along the coast of Chile. The second largest, with a magnitude of 9.4, shook the Alaskan coast in March 1964. Significant damage occurred in Anchorage and the surrounding area, and 131 people died (fig. 4.1).

The big earthquakes happened as seismology was entering a new era. Advances in seismological theory and the advent of computers began making it possible to learn more about earthquakes. Moreover, the newly developing theory of plate tectonics gave seismologists what they had always needed—an explanation of why earthquakes happen. Finally, seismology was becoming a

FIGURE 4.1 Homes in Anchorage damaged by a landslide triggered by the 1964 Alaskan earthquake. (USGS)

tool of the Cold War. Nuclear bomb tests in remote secret areas of the Soviet Union could be detected and studied using the seismic waves they generated. Seismology, traditionally an underfunded, small enterprise, was becoming big science.

The Department of Commerce's Coast and Geodetic Survey ran the government's seismology program. With funding from the Department of Defense, it had begun installing the first worldwide network of standardized seismometers, which would lead to dramatic advances. "And then suddenly," in Charles Richter's words, "the Geological Survey got interested" and succeeded "in practically elbowing the Coast and Geodetic Survey out of the picture." By 1973, the seismic network and National Earthquake Information Center, which locates earthquakes worldwide, were transferred to the USGS.

The USGS and its allies in universities also proposed a major new government program of earthquake studies. It would include scientific study of how earthquakes work and engineering research to make buildings more earthquake resistant. Most dramatically, it would predict earthquakes.

Seismologists had been interested in earthquake prediction for years without success. However, in the late 1960s lab experiments found that just before rocks break, their properties, including the speed of sound waves inside them, changed. If these results from small samples in a lab applied to the messy world of real faults, earthquake prediction could work. Prediction programs began in the U.S., China, Japan, and the Soviet Union. Initial successes were reported. Leading scientists spoke optimistically of the future. Louis Pakiser of the USGS announced that if funding were granted, scientists would "be able to predict earthquakes in five years." California Senator Alan Cranston, prediction's leading political supporter, told reporters "we have the technology to develop a reliable prediction system already at hand."

The prospects for prediction looked different to some of us who were graduate students in the 1970s. We were vocally skeptical, and many of the faculty quietly agreed. Some of the difference was generational. Our senior leaders, from the World War II generation, believed that a powerful government/scientific enterprise could accomplish almost anything. Having watched the new theory of plate tectonics move quickly from vague hypothesis to established fact, the idea that earthquake prediction would do the same seemed natural. In contrast, we students were from the Vietnam generation and had profound suspicion of "the Establishment." To us, the case for earthquake prediction seemed weak, and claims of imminent success sounded like those that victory in Vietnam was "right around the corner." One of Japan's best young seismologists used a similar analogy for his country's expensive but unsuccessful prediction program, saying it was like the way the Japanese army in World War II claimed to be winning despite steady defeats. By cleverness or luck, we proved correct.

Although the prediction bandwagon had started rolling, selling the earthquake program in Washington was tough. Its supporters had different views about how funds should be divided between science and engineering. Other programs were competing for funding. Some in Washington questioned the need for spending large sums on earthquakes given that they are only a minor cause of deaths in the U.S. Despite additional interest generated by the 1971 San Fernando earthquake that killed 65 people in the Los Angeles area, Senator Cranston still lacked the support in 1975 to move the bill starting the program out of committee to a vote. In historian Carl-Henry Geschwind's words, "Advocates of prediction stood in need of a dramatic event to highlight the importance of their work."

At this key time, the USGS dropped a bombshell. In late 1975, USGS researchers reported that a huge area along the San Andreas fault near Palmdale, California, had risen by about 15 inches. Since this part of the fault had ruptured in a major earthquake in 1857, the "Palmdale Bulge" was interpreted as evidence for an impending large earthquake. In his book *California Earthquakes: Science, Risk, and the Politics of Hazard Mitigation,* Geschwind describes how USGS Director Vincent McKelvey expressed his view that "a great earthquake" would occur "in the area presently affected by the . . . 'Palmdale Bulge' . . . possibly within the next decade" that might cause up to 12,000 deaths, 48,000 serious injuries, 40,000 damaged buildings, and up to $25 billion in damage. The California Seismic Safety Commission stated that "the uplift should be considered a possible threat to public safety" and urged immediate action to prepare for a possible disaster. News media joined the cry.

Some of us in graduate school thought the whole thing was silly. At the first seminar I attended in graduate school, I argued with the USGS scientist presenting the bulge results. These came from using surveyor's levels to measure successive positions along lines more than a hundred miles long. Although I had no experience in the matter, I thought that the errors involved in measuring positions from a site on the coast at sea level to Palmdale by a traverse across the San Gabriel Mountains must be too large for meaningful results.

Probably by luck, I was right. Knowledgeable scientists including Dave Jackson of the University of California, Los Angeles and Bill Strange of the National Geodetic Survey had the same criticism. Within a few years, their analyses, in USGS seismologist Susan Hough's words, converted the bulge to "the Palmdale soufflé—flattened almost entirely by careful analysis of data." In hindsight, the bulge incident came from a problem every scientist faces: overinterpreting messy data.

None of this mattered in Washington. Senator Cranston and his allies used the bulge to lobby for increased earthquake funding. The bulge was brought to the attention of Vice President Nelson Rockefeller and other government leaders, who allocated $2.6 million to study it. Soon, the funding dam crumbled. In 1977, Congress passed the Earthquake Hazards Reduction Act. Two years later, the Federal Emergency Management Agency (FEMA) was created and put in charge of coordinating federal earthquake efforts along with the USGS and National Science Foundation (NSF). The program became known as the National Earthquake Hazards Reduction Program (NEHRP), pronounced "knee-herp."

The U.S. government was now solidly in the earthquake business. NSF issued more research grants. The USGS rapidly ramped up its earthquake program, hiring new staff and expanding facilities. It also built a university constituency by awarding contracts to academic researchers working on topics matching its goals.

Soon, however, the prediction program sputtered and then collapsed. Its height— or nadir—came in 1985 when the USGS launched an official earthquake prediction experiment. Part of the San Andreas fault near Parkfield, California, had had moderate (magnitude 6) earthquakes about every 22 years, with the last in 1966 (fig. 4.2). Thus, the USGS predicted at a 95% confidence level that the next Parkfield earthquake would occur within five years of 1988, or before 1993. The USGS National Earthquake Prediction Evaluation Council endorsed the prediction. Equipment was set up to monitor what would happen before and during the earthquake. The *Economist* magazine commented, "Parkfield is geophysics' Waterloo. If the earthquake comes without warnings of any kind, earthquakes are unpredictable and science is defeated. There will be no excuses left, for never has an ambush been more carefully laid."

Exactly that happened. The earthquake didn't occur by 1993, leading *Science* magazine to conclude, "Seismologists' first official earthquake forecast has failed, ushering in an era of heightened uncertainty and more modest ambitions." Although a USGS review committee criticized "the misconception that the experiment has now somehow failed," few seismologists were convinced. USGS wags joked of the $20 million spent on the "Porkfield" experiment. Beyond bad luck, a likely explanation was that the uncertainty in the repeat time had been underestimated. As critics including James Savage, one of the USGS's most respected earthquake scientists, had pointed out soon after the prediction was made, the prediction discounted the fact that the 1934 earthquake didn't fit the pattern well.

When the earthquake finally arrived in 2004, 16 years late, nothing unusual happened before it. At this point, the USGS quietly abandoned official prediction efforts. Few researchers, even those within the USGS, mourned this change because most had long since abandoned prediction studies. As in universities and industry, most were more realistic than their superiors. Some joked that the USGS motto "science for a changing world" was now "science, for a change."

A lot of us, especially outside the USGS, thought the prediction program never had a chance. It counted on measuring something special in the earth

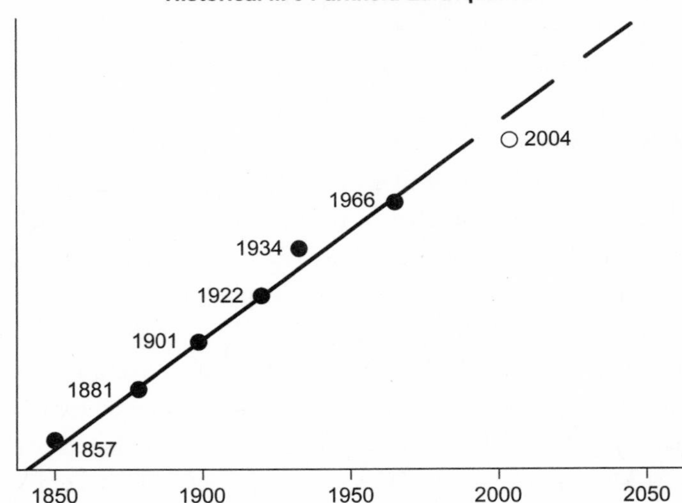

FIGURE 4.2 The Parkfield earthquake predicted to occur within five years of 1988 occurred in 2004. Black dots show the dates of earthquakes before the prediction was made, and the line shows when earthquakes 22 years apart should happen. (After USGS)

that happens before a big earthquake. This hasn't worked yet, despite a lot of looking. Seismologists will keep trying, but at least for now prediction seems far away.

The reason, many of us think, is that tiny earthquakes happen often on a fault, but only a few grow into the rare big ones. If this is right, it will be hard or impossible to predict the big earthquakes because there's nothing special about those tiny earthquakes that by chance grow into big ones. The earth doesn't know which will grow into big earthquakes, and so it can't tell us.

The failure to predict earthquakes seems to be part of a broader pattern, namely that some things are too complicated to predict. This situation, called chaos, happens where some small events grow to have big effects. For example, the flap of a butterfly's wings in Brazil might in theory set off a tornado in Texas. In these cases, it's hard or impossible to predict what will happen. This idea was featured in the book and movie *Jurassic Park*, where small problems grew and caused the whole dinosaur park to collapse.

During the 1970s, the USGS also expanded its interest in New Madrid. This interest had begun in 1904 when geologist Myron Fuller traveled through the area and examined the effects of the earthquakes that were still visible. Fuller made the important observation that there were signs of earlier large earthquakes before those of 1811–1812. He also compiled accounts of the 1811–1812 earthquakes and produced a monograph reporting his studies.

The new USGS program in the central U.S. started off well in 1973. USGS and university scientists investigated geologic structures to learn how they might be related to the earthquakes. They also mapped how the ground had deformed over long time intervals and found geological evidence for older earthquakes.

However, a subtle change happened. The USGS opened a field office in Memphis and moved from investigating what was going on to deciding that there was a high hazard and promoting it. Some of the scientific results were good. However, all data were interpreted using the preconceived high hazard idea. Any observations that could in any way be interpreted as favoring high hazard were welcome; data that didn't fit got tuned out. It's like George Bernard Shaw's play *Saint Joan,* in which the definition of a miracle is something that confirms the faith.

A major move in this direction came from a USGS program to make new earthquake hazard maps. The mapping results, announced in 1996, were stunning. In previous maps, the predicted hazard in the New Madrid zone was much lower than in California, but the new maps made it as high or even higher.

The question was whether to believe the maps. Although few scientists inside or outside the USGS knew what new assumptions in the computer program produced the change, many thought it didn't make sense. After all, every few years an earthquake in California is big enough to cause serious damage, but that hadn't happened in the Midwest in more than 100 years. Zhenming Wang and colleagues at the Kentucky Geological Survey started questioning the science behind the maps, and USGS researchers in California joked that their Memphis colleagues were visiting to escape the danger back home.

Earthquake Engineers

The new maps were developed with input from earthquake engineers, who play a key role in our story. Earthquake engineers are crucial in earthquake safety because they design buildings to survive shaking by seismic waves. They face a potential conflict of interest because they have a major influence in writing the legal codes that specify how buildings should be built—and then are in the business of meeting those requirements. This conflict hasn't been a major problem in California because damaging earthquakes are common enough that there's general agreement about how strongly buildings should be built.

Based on the California experience, earthquake engineers assumed that seismologists understood the earthquake hazard in the Midwest so they could design buildings to meet it. Building to California standards in the central U.S. seemed reasonable to many of them. However, from a seismological view, it didn't seem sensible. The Midwest situation is very different from California because damaging earthquakes are extremely uncommon. The largest in the past 100 years, a magnitude 5.5 earthquake that occurred in southern Illinois in 1968, caused no fatalities. Only minor damage occurred: fallen bricks from chimneys, broken windows, toppled television aerials, and cracked or fallen brick and plaster (fig. 4.3). Thus, there's some hazard but much less than California's. The question that needs careful analysis is what level of construction makes sense in the Midwest.

In this case, scientists and engineers have different views because of their cultural differences. People become geologists because they're fascinated by

FIGURE 4.3 Only minor damage resulted from the 1968 magnitude 5.5 southern Illinois earthquake, the largest in the Midwest in the past 100 years. (St. Louis University)

the mysteries of nature. Despite our technology and tools, we're passive observers of these complexities. We recognize them and are comfortable admitting what we don't know and that it will be tough, slow going to do better. On the other hand, we expect that our views will change as we get new data or ideas. Engineers are activists who shape the man-made world. They want geological problems simplified so they can get to work. Where we see complexity, they see quibbles.

These differences are shown by a conversation between Michael Wysession, a seismologist at Washington University in St. Louis who is a former student of mine, and an earthquake engineer. The engineer criticized me for arguing that the earthquake hazard was less than claimed and was surprised by Michael's response that the goal of seismology was to find out what the hazard actually was, not to promote a high one.

As a result, seismologists joke about engineers, and engineers probably do the reverse. USGS seismologists used to say that they worked for a group studying three plagues afflicting mankind—the Office of Earthquakes, Volcanoes, and Engineering.

Emergency Managers

Ideally, emergency management agencies would give the public calm, reasoned assessments of potential hazards and carefully thought out policies to address them. Unfortunately, this hasn't happened for Midwest earthquakes.

FEMA has played the most active role. Part of the Department of Homeland Security, FEMA has about 2,500 employees and a $6.5 billion annual budget. Prior to August 2005, when Hurricane Katrina struck New Orleans and its surroundings, many Americans knew little of this organization. Its botched response to Katrina made FEMA famous. Director Michael Brown, who took the job after serving as a commissioner of the International Arabian Horse Association, knew less about what was going on than millions of TV viewers. When informed that thousands of people were trapped with little food and water in dangerous conditions in the city's convention center, Brown responded, "Thanks for the update. Anything specific I need to do or tweak?" Although he didn't go to the city or send help, he took care to be photographed with sleeves rolled up as a sign of activity. When firefighters from across the country responded with their rescue gear, ready to go, FEMA sent them to Atlanta for lectures on equal opportunity and sexual harassment.

FEMA did no better during the reconstruction phase. It housed people in trailers that made their inhabitants sick from poison fumes. Its bureaucracy became a national laughingstock, illustrated by the famous case of restocking New Orleans' aquarium. FEMA wanted the aquarium to buy fish from commercial vendors and initially refused to pay when the aquarium staff caught the fish themselves, saving taxpayers more than half a million dollars. Simon Winchester, who wrote a history of the 1906 San Francisco earthquake, noted that the government responded far better to that event, with steam trains, horse-drawn wagons, and no emergency management agencies, than it did to Katrina 100 years later with designated agencies, jet aircraft, helicopters, and television.

FEMA's involvement with Midwest earthquake hazards has been no better. The agency is the primary force behind the pressure to build buildings in the Midwest to the same earthquake standard as in California. This position came out of a complicated bureaucratic process involving FEMA, USGS, and earthquake engineers working through a FEMA-sponsored group called the Building Seismic Safety Council. All of this resulted in a new tough building code that FEMA is pressuring states and communities to adopt. Although FEMA wants immediate action, the science, economics, and logic behind their plan are weak. Still, as Julius Caesar said, men readily believe what they want to believe.

Some of the loudest support comes from state emergency managers who form the Central U.S. Earthquake Consortium, CUSEC. The group says its mission is "the reduction of deaths, injuries, property damage and economic losses resulting from earthquakes in the Central U.S." and seeks "to increase national readiness for a catastrophic earthquake in the New Madrid Seismic Zone." The problem is that they start off assuming that disaster looms rather than with the question of whether it really does. That's putting the cart before the horse; it's hard to make sensible plans until we figure out what the hazard is.

As a result, much of what CUSEC says is silly, but it sounds scary and gets attention. In 2000, CUSEC Executive Director James Wilkinson announced, "Seismologists have predicted a 40–60% chance of a devastating earthquake in the New Madrid seismic zone in the next 10 years. Those odds jump to

90% over the next 50 years. The potential magnitude of a catastrophic New Madrid quake dictates that we approach the preparedness on a regional basis." The "catastrophic" and "devastating" earthquakes that he was talking about are magnitude 6, much smaller than the 1811–1812 earthquakes. To make things worse, there's no such thing as "the" probability of an earthquake, because it can be calculated in many ways. Because a magnitude 6 earthquake happens about every 175 years in the area, the particular probability he was quoting in 10 years is really 10/175 or 6%—not 60%. Wilkinson was also quoted as saying that the question was "not if but when" large earthquakes like those of 1811–1812 would strike again. In reality, geologists don't know whether, where, or when such earthquakes might occur. Unfortunately, CUSEC's statements don't acknowledge the complications of the real world.

Again, the different views reflect different cultures. Emergency managers have the tough job of preparing for disasters. They live in the political world and need to preserve and grow their agencies in typical years when no disasters happen. This situation can create a tendency to warn about possible hazards, however unlikely, and urge immediate action. They have no incentive not to recommend spending major resources preparing for an unlikely disaster, but their jobs are at stake if a disaster occurs for which they weren't ready.

In many natural hazard situations, including earthquakes in California, geologists and emergency managers largely agree. We accept their need to present hazards simply to the public. However, for New Madrid, the emergency managers want to act aggressively before science understands what the hazard is.

The idea of making buildings as earthquake resistant as in California sounded good at first, and some states agreed. However, no one had mentioned a problem—the high cost. By any measure, it would be billions of dollars over hundreds of years. It turned out that FEMA and the earthquake engineers hadn't studied the economics of their proposed code. No one had any idea how the costs and benefits compared. FEMA was like an insurance salesman promoting a new policy without any numbers. State and local governments, business groups, and even some engineers began questioning whether the new code made sense. Wouldn't it be more sensible to adopt more modest and less expensive standards, at least until we figure out what the hazard is?

That's certainly my view. I see the earthquake hazard as a long-term problem that should be studied thoroughly and carefully, especially because major earthquakes are unlikely to strike for hundreds or even thousands of years.

Rushing to spend billions of dollars is likely to be wasteful because as the data get better, New Madrid looks less dangerous.

Zombie Science

Although the media loved the drumbeat of impending doom, many midwestern seismologists were embarrassed by it. Increasingly, earth scientists decided not to get involved in New Madrid studies. There was plenty of interesting science to do elsewhere without tangling with what some called "the New Madrid Mafia." Even Nuttli's successors at St. Louis University put less effort into New Madrid research. A common view was summarized by Tom Jordan, director of the Southern California Earthquake Center, who told me, "I wouldn't touch New Madrid with a 10-foot pole."

Fortunately, when things get out of hand, science has a way of correcting itself. Although this correction started with our GPS measurements, we hadn't planned it. When we started the GPS program, our interests were focused on the scientific question of what causes earthquakes in the middle of the continent. I didn't think the hazard was very high but hadn't thought about why it was claimed to be so high. I'd never thought about building codes or hazard policy.

All that changed in 1997 when we got our third set of GPS measurements showing no motion within the seismic zone. The simplest interpretation was that no large earthquake is on the way. The fault is shutting down for a while, so all we'd expect is small earthquakes. Thus any future 1811–1812-style earthquake is very far in the future and the hazard is nowhere near as high as was being claimed.

At first, this idea surprised many geologists. Although there had been no way of telling, we'd thought that a big New Madrid earthquake was on the way within the next few hundred years. Suddenly there was GPS data that could show if a big earthquake was coming, and there was no sign that one was.

Because I wasn't committed to the idea of an upcoming large earthquake, giving up on it was easy. Many geologists have the same reaction, but others have more difficulty. They recognize the problem the GPS data pose but would like to fit them into the idea of a big earthquake on the way.

That gets harder year after year as the GPS data showing little or no motion get better. To say that a big earthquake is coming calls for special arguments in which New Madrid behaves differently from any other earthquake zone that we know. That's not impossible, but there's no reason to expect it.

The situation is like someone who's sure that he has a broken bone, but X-rays don't show one. He asks for a CAT scan, which also doesn't show a break. An MRI also comes up negative. It's still possible that there's a break that for some unknown reason doesn't show up, but as the images get better it's increasingly likely that there simply isn't a break.

It's interesting to think about how to resolve this question. If the ground starts moving or a big earthquake happens, then the fault hasn't shut down. Conversely, my question to anyone who expects a major earthquake soon is: What data would persuade you otherwise? Fifty years of GPS data showing no motion? Two hundred years? If 500 years from now the GPS data still show no motion and the major earthquake hasn't happened, would they accept that the fault system has shut down? Or would they say that the earthquake is coming, just delayed?

If no data could change their view, then it's their belief. Beliefs are ideas that data can't change, like that the world will end on December 21, 2012, at the end of the ancient Mayan calendar. When that doesn't happen, believers will simply pick a later date, just the way they did when the world didn't end as predicted at the end of the millennium in 2000. Beliefs are fine, but they aren't science.

I'm sometimes asked whether those who still argue that New Madrid's earthquake danger is as high as California's really believe that. I don't know, of course. I suspect that most recognize the problems with their argument. However, although scientists can change their views in response to data, it's hard for government agencies to change without losing face.

The rest of this book explains how the idea of a huge hazard grew and is now dying. I'll go through the science, explain what went wrong in the studies that argued for a high hazard, and give what I think is a more sensible picture. You'll see why I think the idea that there's a huge earthquake hazard in the Midwest is a case of "zombie science"—a theory that's scientifically dead but is still walking around as it fades away. Unfortunately, the old idea is still causing needless fear and wasting a lot of money.

Chapter 5

Earthquake!

We have seen a statement made by a couple of gentlemen just from New Madrid, which says that that place is much torn to pieces by the late earthquake.

—*New York Post*, March 11, 1812

River World

To most Americans in 1811, the Mississippi valley was a strange and remote world. The new nation—the Constitution had been ratified only 24 years earlier—stretched along the Atlantic coast. Its economy, culture, and politics centered on seaports surrounded by prosperous farms. These cities, like Philadelphia, Boston, New York, and Charleston, had fine public buildings, colleges, and newspapers that rivaled Europe's. Philadelphia was the second largest English-speaking city in the world after London. American ships traded worldwide.

The frontier lands west of the Appalachian Mountains were very different. Since the American Revolution, tough individualists had been making their way west to settle these harsh and dangerous lands. Most settled along the rivers that were highways through the wilderness. One Mississippi River town, New Madrid, gave its name to the earthquakes making our story.

To understand New Madrid, we need to understand the Mississippi River. By any measure, it's big. It's the fourth-longest river in the world and is tenth in the amount of water it carries. It drains the middle third of the U.S., almost all of the area between the Appalachian Mountains and the Rocky Mountains. From its source in Lake Itasca in Minnesota to its mouth at the Gulf of Mexico, it's 2,320 miles long. Along the way, major tributaries like the Wisconsin, Illinois, Missouri, Ohio, and Arkansas rivers feed into it.

Many large cities grew up because of the river. New Orleans is the great river port, St. Louis is where the Mississippi and Missouri meet, and Chicago is the place where boats from the Great Lakes—which connect to the Atlantic—can enter the Illinois River and then the Mississippi.

The Mississippi shapes both the geography and human lives along it. It's a powerful force. As Mark Twain, who piloted steamboats on it in the 1850s, described, "It is not like most rivers beautiful to the sight, bestowing fertility in its course; not one that the eye loves to dwell upon as it sweeps along, nor can you wander upon its banks, or trust yourself without danger to its stream. It is a furious, rapid, desolating torrent, loaded with alluvial soil." The famous muddy waters carry over 400,000 tons of sediment every day on average. Some of the mud makes its way to the Gulf of Mexico, but some doesn't get that far.

Where the Mississippi—or any other river—passes through a bend, a funny thing happens. As canoeists and kayakers learn, the fast way around the bend is the long way, because water on the outside of the bend flows faster. The faster moving water on the outside erodes the riverbank, while the slower moving water near the inside deposits more sediment there. Over time, these bends, called meanders, grow bigger. However, as usual, nature has a way of stopping things from getting out of hand. When a meander gets too big, its narrow neck is eventually cut off from both sides, straightening the river and leaving an "oxbow" lake behind.

Meanders are normal features of mature rivers that flow across broad valleys, leaving a broad flood plain full of thick, young, loose sediments. Thus, like most broad river valleys around the world, the Mississippi valley is great for farming. What's unusual is that unlike most of these valleys, parts of the Mississippi valley have big earthquakes.

Geologists view nature and society as closely connected. The New Madrid earthquakes and the Mississippi River are a great example. The river is where it is in part because of the faults deep below the sediment that cause the earthquakes, and earthquakes change the river's course. The people are there largely because of the river and its thick sediment deposits. The sediments might even be part of the reason the earthquakes happen.

The sediments both hurt and help us in studying the earthquakes. They make earthquakes harder to study because they hide the faults below. On the other hand, they give a valuable history of earthquakes going back before written records. That's because the sediments include layers of watery sand

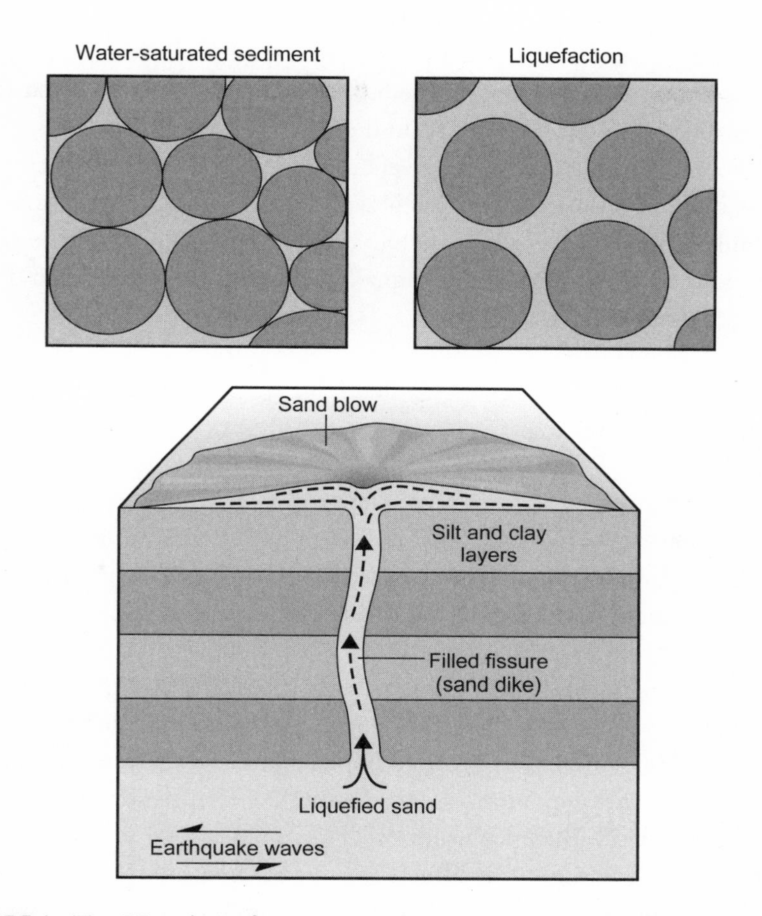

FIGURE 5.1 *Top,* How liquefaction occurs; *Bottom,* Schematic diagram of how sand blows form. (USGS)

that have been buried under layers of more muddy soil. Earthquake shaking can make the sand liquefy. Grains lose contact with each other, so the sand acts like a liquid and erupts through the overlying soils to form sand blows (fig. 5.1). Sand blows from the 1811–1812 earthquakes are easily seen from the air as sandy white patches when fields are bare or as withered yellow patches in a sea of green during growing season.

Sand blows have also been found from earlier large earthquakes around the years 900 and 1450 that were probably like the sequence of earthquakes in 1811–1812. These dates come from carbon dating of plant material in the sand blows. This dating uses the fact that air contains a small amount of radioactive carbon, which ends up in growing plants. Once plants die, the amount of radioactive carbon in them decreases with time, so measuring it tells how long it's been since the plant died.

Although the thick sediment layers make the river valley attractive for people, they're a problem during earthquakes. They're easy to shake, so buildings built on them face greater risk than buildings on solid rock. In addition, if the ground liquefies in a big earthquake, buildings can topple even if they survive the shaking.

Life on the Mississippi

The earth and people have different time scales. Earthquakes happen on a long and slow geological time scale, so they rarely affect humans during our short lives. Thus, although European settlers in the Mississippi valley probably didn't know that large earthquakes could happen there, their Indian neighbors did.

As far as we know, the first humans hit by one of these large earthquakes were Indians whose culture flourished in the Mississippi valley and nearby from about 800–1500 A.D. They lived in prosperous communities with economies based on fishing, hunting, and farming the river floodplains for corn, beans, and squash. Their towns included large flat-topped mounds like the spectacular ones at Cahokia, Illinois, east of St. Louis. At its peak around 1150 A.D., Cahokia is thought to have had at least 20,000 people, more than lived in London.

In 1846, Charles Lyell, one of the founders of geology, inspected the effects of the 1811–1812 earthquakes and was told of Indian legends describing an earlier large earthquake. Although details of the legends have been lost, Indian artifacts are found in sand blows from past earthquakes, showing that the stories were rooted in fact. It's been suggested that the New Madrid earthquakes that occurred around 900 A.D. may have had a cultural effect. The coincidence of this earthquake sequence and the emergence of the Mississippian Indian culture suggest that the former may have provided the impetus for the latter.

The first written records of New Madrid earthquakes are from the ones that struck over the winter of 1811–1812. At the time, the Mississippi river valley was the frontier. Land to the west, which had been French Louisiana, became Spanish in 1763 under the terms of the treaty that ended the French and Indian War. By controlling New Orleans, Spain had closed the Mississippi to American traffic. Any viable Mississippi valley settlement required Spanish approval.

New Madrid, which claims to be the oldest city west of the Mississippi, was settled in 1789 by a group of 70 Americans led by American Revolution Colonel George Morgan. The settlers came down the Ohio River in four boats from Fort Pitt, now modern-day Pittsburgh, to the Mississippi.

The town was named New Madrid in hope of pleasing the Spanish, from whom Morgan planned to obtain a large land grant. He chose the site at the major river bend about 70 miles south of where the Ohio meets the Mississippi. Morgan laid out a street plan for a city including parks, churches, and schools that would be the primary transfer point through which American settlers would connect with river trade.

The plan encountered political and geological difficulties. Esteban Miro, the Spanish territorial governor, refused to grant Morgan the governing power and right to sell land that he wanted. Morgan thus lost interest in the colony and left. The Spanish then established a fort at New Madrid and many of the original settlers left. To make things worse, the Mississippi soon showed its power. The high bank on which the town was built eroded rapidly. Within five years, the fort built 600 yards from the river was collapsing. The river didn't need help from an earthquake to bring down structures along the riverbank, or the bank itself. As Mark Twain would later describe, "Pouring its impetuous waters through wild tracks . . . it sweeps down whole forests in its course, which disappear in tumultuous confusion, whirled away by the stream now loaded with the masses of soil."

Meanwhile, colonial politics on the frontier continued. In 1800, under pressure from Napoleon, Spain returned Louisiana to France. This situation alarmed President Thomas Jefferson, who feared that Napoleon planned a great French empire that would block America's westward expansion. However, after losing the Caribbean island of Hispaniola to a slave rebellion and the beginning of a war with England, Napoleon had little use for Louisiana. He offered to sell it, and Jefferson accepted. This proposal triggered strong opposition by Alexander Hamilton's Federalist party, in part because of the fear that the country would be split by tension between the established east and the rising west. However, in 1803 the U.S. Senate ratified the Louisiana Purchase treaty. The purchase doubled the size of the nation and opened the Mississippi, all for about $15 million.

By 1811, river traffic had grown as the river became the major transportation route for settlements west of the Appalachians. Most people in the area lived in small towns along the river. Even St. Louis was small, and the city of Memphis didn't exist yet. Although the New Madrid area had about 3,000 residents, the town's situation was precarious. In the words of historian James Penick, "Its founder had hoped they would stand astride the wealth of a continent, but 1811 found the people of New Madrid locked in an endless, hopeless struggle with the unrelenting river that was to have been their highway to greatness."

More generally, in Penick's words, "the year 1811 had little to recommend it." Tension with England had been building for years, and war seemed forthcoming. The country was bitterly divided. "War hawks" from the south and west favored war to expand the country while New Englanders and New Yorkers were opposed. The U.S. was already fighting frontier Indians backed by England. Despite the November 1811 U.S. defeat of a coalition of tribes led by the Shawnee leader Tecumseh at the Battle of Tippecanoe, the threat remained. Tecumseh is said to have told the tribes that he would stamp his foot and the Great Spirit would shake the earth all along the river. Although this story is widely circulated, including in a USGS publication "Tecumseh's Prophecy," it seems to be part of the legend surrounding New Madrid earthquakes. Analyses of the historical record by Penick and USGS seismologist Susan Hough conclude that there probably wasn't a prophecy. Instead, addressing a gathering after the earthquake, Tecumseh pointed to it as divine support for his cause. "Brothers," he told his followers, "the Great Spirit is angry with our enemies. He speaks in thunder, and the earth swallows up villages." This wasn't a prophecy, but it was a good description of what had happened.

The Ground Shakes

The bad year of 1811 got even worse just after 2 AM on December 16, when the New Madrid fault system came to life. These faults are huge cracks in the solid rocks far below the sediments of the river valley. For hundreds of years, geological forces had been storing up energy in the rocks, much like the way squeezing a spring stores energy in it. Eventually, the stored energy overcame the friction between rocks on the two sides of the faults. The rocks started to slide, and the energy stored over years came out in seconds, shaking the ground above.

Even today, when scientists understand a lot about earthquakes, they're terrifying. In 1811, no one had any idea what was happening. Still, many people had the presence of mind to give vivid accounts of what they experienced and how they felt.

A famous description comes from a resident of the town of New Madrid:

> About two o'clock this morning we were awakened by a most tremulous noise, while the house danced about, and seemed as if it would fall on our heads. I soon conjectured the cause of our trouble, and cried out that it was an Earthquake, and for the family to leave the house, which we found very difficult to do, owing to its rolling and jostling about. The shock was soon over, and no injury was sustained, except the loss of the chimney.

Adding to the surreal scene, "a vapour seemed to impregnate the atmosphere, had a disagreeable smell, and produced a difficulty of respiration."

During the night "we had had eight more shocks, none of them so violent as the first." However in the morning, "we had another shock. This one was the most violent we have yet had."

Many accounts come from riverboats, the heavy trucks of their day that transported most goods. Because the river was too dangerous to navigate at night, they moored by the shore or at islands, often in groups. Scottish naturalist John Bradbury, who was studying North American plants, was sleeping on one moored near present-day Memphis. Suddenly, he recounted, "I was awakened by a most tremendous noise accompanied by so violent agitation of the boat that it appeared in danger of unsettling." Once on deck, he could "distinctly see the river as if agitated by a storm and although the noise was inconceivably loud and terrific I could distinctly hear the sound of falling trees and the screaming of the wild fowl on the river, but found that the boat was still safe at its moorings."

Earthquakes went on all night, and by daylight "a shock occurred nearly equal to the first in violence." In many places the riverbanks collapsed and put the boats in danger from falling trees that "rushed from the forest, precipitating themselves into the water with a force sufficient to have dashed us into a thousand atoms." Boats got underway as soon as it was light and traveled downriver while watching the banks continue to collapse and encountering the large waves that resulted. Some boats were "cast high and dry on the shores." This happened in particular near New Madrid, leaving people there with supplies that probably helped them through the events that followed.

An even more horrific picture comes from an account by John Walker, who was camped by a small lake near Little Prairie, near present-day Caruthersville, Missouri. He and a friend were "were awakened by a noise like distant thunder, and a trembling of the earth, which brought us both to our feet." The dawn aftershock was much scarier:

> It was awful! Like the other—first, a noise in the west, like heavy thunder, then the earth came rolling towards us, like a wave on the ocean, in long seas, not less than fifteen feet high. The tops of the largest sycamores bending as if they were coming to the ground—again, one rises as if it were to reinstate, and bending the other way, it breaks in twain, and comes to the ground with a tremendous crash. Trees were falling in every direction—some torn up by their roots, others breaking off above the ground, and limbs and branches of all sizes flying about us, the earth opening, as it were, to receive us, in gaps sometimes fifteen feet wide—then it would close with the wave. The water of our little lake was fast emptying itself in these openings, and as soon as they would close, it would spout high in the air.

By the time the shaking ended, "The whole forest seemed as if an awful hurricane had completely destroyed it. The soft alluvial earth was opened in many rents of great depth, in which our little lake had completely lost itself."

The ground in the town of Little Prairie was equally disrupted. "There was not perhaps a square acre of ground unbroken in the neighborhood, and in about fifteen minutes after the shock the water rose round them waist deep." Not surprisingly, almost the entire population, about 100 people, left town and walked the 30 miles north to New Madrid because they heard that "the upper country was not damaged."

These accounts of ground damage are believable. Almost a century later, geologist Myron Fuller visited the area and could still see effects of the earthquakes, including broken ground and trees that were tilted and later grew vertically (fig. 5.2).

FIGURE 5.2 Effects of the 1811 and 1812 earthquakes photographed by Myron Fuller in 1904. (USGS)

The earthquakes went on and on. Most were small, but one on the morning of January 23, 1812, was large enough to disrupt riverbanks and create more sand blows.

The worst shaking came early on the morning of February 7, 1812 (fig. 5.3). In the words of New Madrid resident Eliza Bryan, "a concussion took place so much more violent than those preceding it that it is denominated the 'hard shock.'" The town's houses, which sustained damage like broken chimneys in the previous earthquakes but had not collapsed, were "all thrown down." There "was scarcely a house left entire—some wholly prostrated, others unroofed and not a chimney standing." This was too much, even for these tough folk who had endured weeks of earthquakes. They "fled in terror from their falling dwellings."

Until this shock, New Madrid itself had been spared significant ground damage. Now, Bryan told of widespread cracking and sand blows. "The earth was horribly torn to pieces. The surface of hundreds of acres was, from time to time, covered over of various depths of the sand which issued from the features . . . some of which closed up immediately after they had vomited forth their sand and water."

What happened to the river in this "hard shock" (fig. 5.4) became the most famous feature of the New Madrid earthquakes. Bryan wrote that

> At first the Mississippi seemed to recede from its banks and its waters gathering like a mountain, leaving for a moment many boats . . . on the bare sand,

FIGURE 5.3 Classic woodcut portraying "the great earthquake at New Madrid." (Image courtesy of the State Historical Society of Missouri)

in which time the poor sailors made their escape from them. It then rising fifteen or twenty feet perpendicularly and expanding as it were at the same moment, the banks were overflowed with a retrograde current, rapid as a torrent—the boats which before had been left on the sand were now torn from their moorings and suddenly driven up a little creek . . . The river fall-

FIGURE 5.4 Drawing of effects of the 1812 earthquake on Mississippi riverboats. (Image courtesy of the State Historical Society of Missouri)

> ing immediately, as rapid as it had risen, receded within its banks again with such violence that it took with it whole groves of young cottonwood trees . . . A great many fish were left on the banks, being unable to keep pace with the water. The river was literally covered with the wrecks of boats.

Accounts from boat crews show that the river was disturbed over a large area, not just at New Madrid. Mathias Speed's two boats had moored for the night at a sandbar 20 miles upriver from New Madrid. At 3 AM, they were "awakened by the violent agitation of the boat, attended by a noise more tremendous and terrific than I can describe or anyone conceive." Because the sandbar was sinking, they cut loose and rowed into the middle of the river, which got them "out of danger from the trees, which were falling in from the banks—the swells in the river were so great as to threaten the sinking of the boat every moment." At daylight, Speed was surprised to discover that the boat had moved "only four miles down the river—from which circumstance and from that of an immense quantity of water rushing into the river from the woods—it is evident that the earth at this place or below had been raised so high as to stop the progress of the river and cause it to overflow its banks." Passing New Madrid, "we were affrighted with the appearance of a dreadful rapid or falls in the river just below us, we were so far in the suck that it was impossible to land—all hope of surviving was now lost and certain destruction appeared to await us. We passed the rapids without injury, both boats being still lashed together." Speed learned from a boat coming upriver that there was a second waterfall about seven miles downriver.

A summary of accounts from different boats gives an overview of what happened:

> The current of the Mississippi . . . was driven back upon its source with the greatest velocity for several hours in consequence of the elevation of its bed. But this noble river was not to be stayed in its course. Its accumulated waters came booming on, and o'er topping the barrier thus suddenly raised, carried everything before them with resistless power. Boats, then floating on its surface, shot down the declivity like an arrow from a bow, amid roaring billows and wild commotion.

Although the reverse current only lasted for a few hours, the new waterfalls survived for several days.

Using the Clues

The historical accounts give a vivid feel for people's experiences during the earthquakes. Naturally, they don't tell us everything we'd like to know. For example, it's not clear if anyone was killed, though the accounts show that there couldn't have been many deaths.

Two important questions scientists ask about earthquakes are which faults moved and how they moved. For modern earthquakes, researchers learn the answers using seismograms, but the seismometer was only invented about one hundred years ago. Without seismograms, you'd think the historical accounts couldn't tell much, but these accounts are combined with additional information from the smaller earthquakes that continue in the area today. Most of these seem to be aftershocks that happen on or near the faults that moved in 1811–1812. Although these faults are buried deep under the sediment, they can be mapped using the locations of the smaller earthquakes.

The recent earthquakes show at least three major faults (fig. 5.5). Although the New Madrid seismic zone is named after the town of New Madrid, it

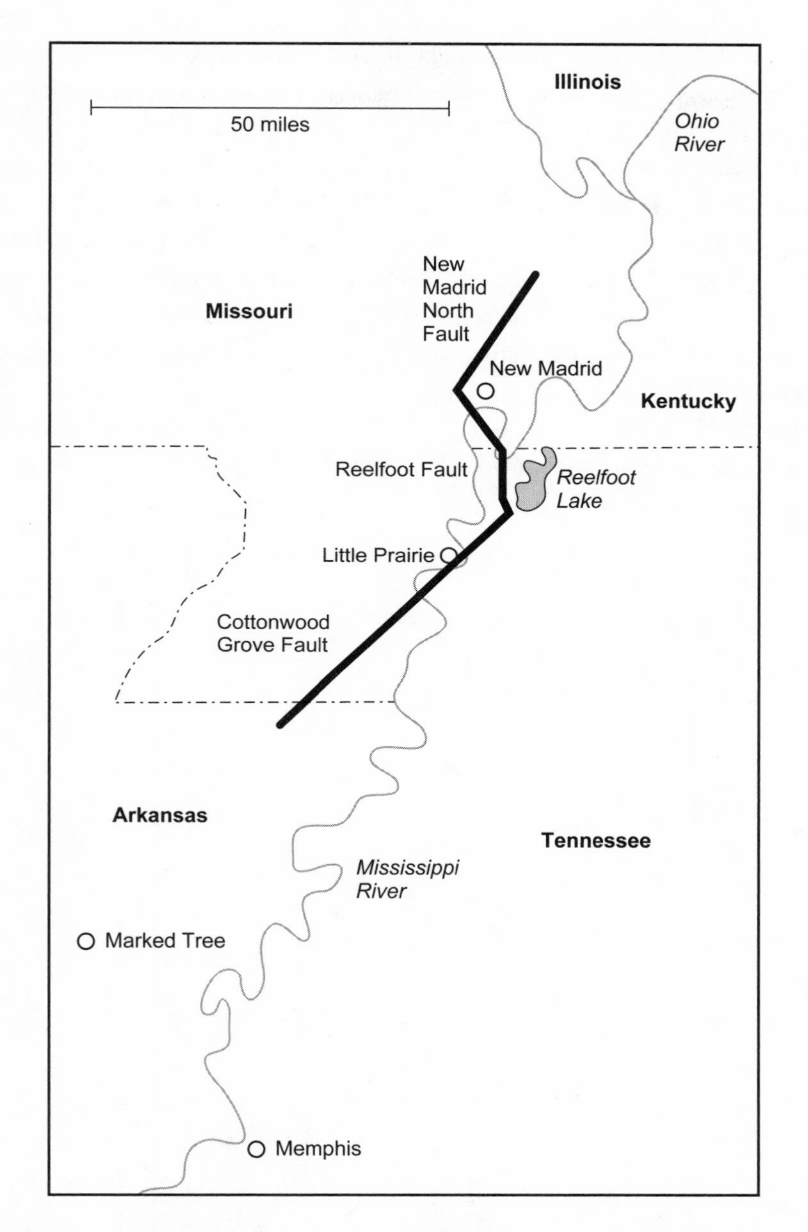

FIGURE 5.5 Map showing New Madrid and surroundings. In 1811, the states had not been established, the city of Memphis did not exist, and no one knew about faults. (After Sieh and LeVay, 1998)

covers a large area. The main active faults start north of New Madrid, extend the length of the Missouri Bootheel, and continue south into Arkansas and east into Kentucky and Tennessee.

The accounts of the 1811–1812 earthquakes give insight about where in the fault zone the quakes occurred. That's because how strongly the ground shakes in an earthquake depends on three things. The first is the magnitude of the earthquake, or how big it is. The second is how far away it is, because seismic waves attenuate—lose some of their energy—as they travel through the earth. The third is the nature of the ground, because soft ground shakes much more.

The first December earthquake probably happened on the southern fault, the Cottonwood Grove fault. Because accounts from different places along the

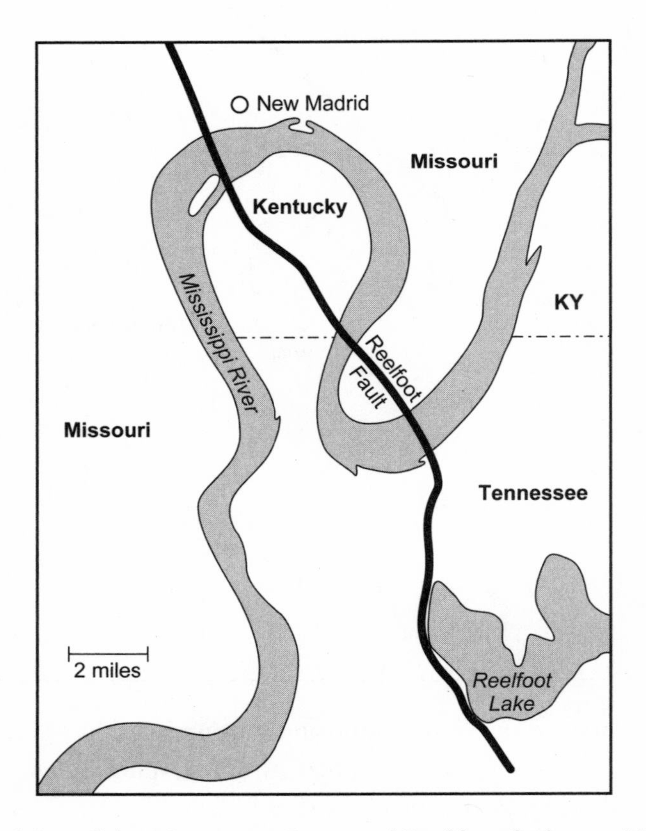

FIGURE 5.6 Map of the Mississippi River and Reelfoot fault near New Madrid. (After Sieh and LeVay, 1998)

fault describe about the same level of shaking, it seems that most of this fault moved in the earthquake.

The earthquake that followed in the morning is thought to have been on the east end of the Reelfoot fault, in the middle part of the seismic zone. Accounts of the two quakes from locations far from the faults tell us that the first shock was the larger of the two. However, because Little Prairie was closer to the second shock, the shaking there was much greater the second time.

Figuring out where the January 26 earthquake happened is tough, but it's thought to have been on the North New Madrid fault. Earthquake researchers are surest of the February 7 "hard shock," which was the largest of the earthquakes. From the descriptions of what happened at New Madrid and on the Mississippi River, the earthquake seems to have been on the Reelfoot fault, close to New Madrid. Vertical motion on the fault raised the ground on the southwest side by about 10 feet relative to the northeast side. This motion created Reelfoot Lake, now a scenic Tennessee state park. At the three places where the fault crosses the river (fig. 5.6), this motion created natural dams on the riverbed that disrupted the flow until the current cleared them away.

The details of what happened aren't clear. In particular, there's been a lot of skepticism about the accounts of the Mississippi flowing backward. This might have been an illusion from the water flowing over the riverbanks. However, those reporting it were experienced boatmen who knew currents. Their accounts remind me of the dangerous situation that kayakers and canoeists encounter when a river flows over what's called a low-head dam (fig. 5.7). Water falling over the dam collides with water below the dam and

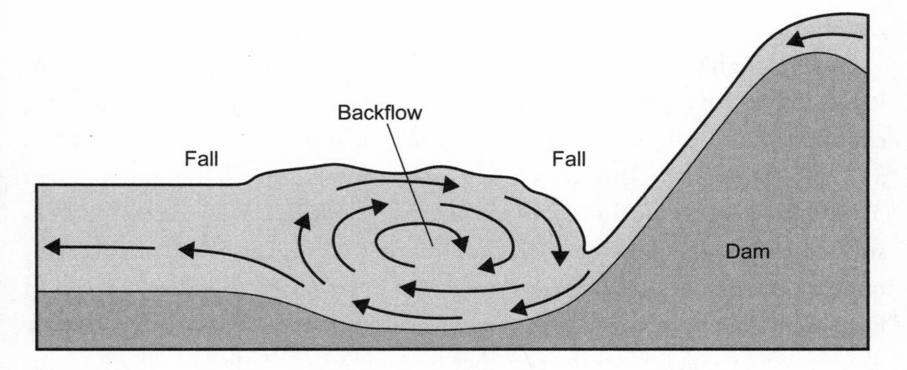

FIGURE 5.7 Water flow near a low-head dam.

so circulates upward and backward. The flow creates a zone where water near the surface flows backward, with waterfalls on both its upstream and downstream sides. I wonder if the Mississippi boatmen encountered a much bigger and more complicated version, with backflow downriver from the first natural dams and slower currents extending upriver. Because the fault crosses the river in three places, the falls and complicated currents would have given the boatmen a scary ride.

More can be learned about the earthquakes by adding reports from places farther from the faults. Shaking was felt over much of the eastern U.S. because the cold, strong rocks of the central and eastern U.S, transmit seismic waves well. However, the historical accounts show that damage was much less than popularly believed today. That's because we often see maps showing the areas of shaking, but shaking gets smaller at longer distances from an earthquake and it takes significant shaking to do much damage.

The map in figure 5.8, compiled by Susan Hough for the first major New Madrid shock, shows that damage was minor except close to New Madrid. About 100 miles from New Madrid, in Ste. Genevieve, Missouri, the shaking caused no major damage. A brick home that survived undamaged can be seen today, which is interesting because brick construction is especially vulnerable to earthquake shaking. In St. Louis, about 150 miles away, a newspaper reported that "No lives have been lost, or has the houses sustained much injury. A few chimneys have been thrown down." Similar minor damage with "no real injury" occurred in Nashville, 160 miles away, as well as in Louisville, Natchez, and Vincennes. No damage occurred in Fort Wayne, Wheeling, Asheville, Brownsville, Norfolk, or Detroit. In the Kentucky hills south of Cincinnati, many families slept through the shock.

Both Hough's and earlier analyses of the historical accounts debunk some legends that have grown up. They show that the earthquakes didn't ring church bells in Boston, contrary to a legend that appears in books, media, and a USGS "Fact Sheet." This story is common; a Google search found more than 32,000 references, including the online encyclopedia Wikipedia. The bells story gives a reality check on sources of information about New Madrid earthquakes because ones that repeat it might be questionable in other aspects. In fact, there are no accounts that anyone in Boston even felt the earthquakes. The story may come from the fact that bells rang in Charleston, South Carolina, which is much closer to New Madrid, and got confused with the Charles-

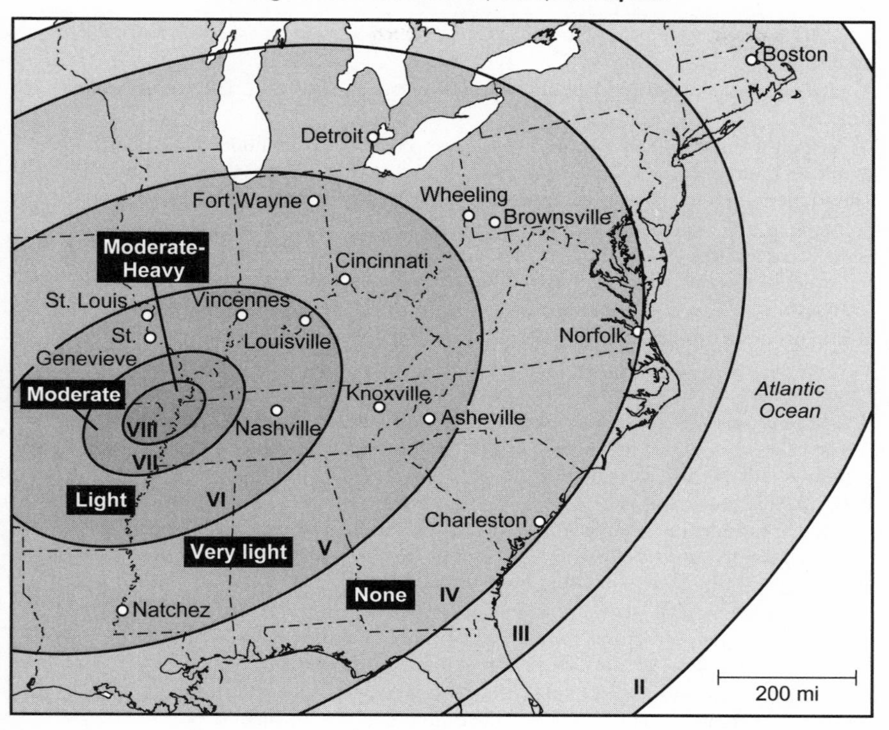

FIGURE 5.8 Intensity map for the first of the three major New Madrid shocks. Boxes label different damage zones corresponding to intensity contours. Some of the sites from which we have reports are shown. (After Hough et al., 2000)

town area of Boston. There's also no historical support for the common stories of damage in Washington, D.C., and chimneys falling in Maine.

Intensity maps let seismologists infer the size, or magnitude, of the earthquakes. Today, earthquake magnitudes are measured from seismograms, as we'll discuss in Chapter 8. Magnitudes for earthquakes that occurred before the seismometer was invented can be inferred by mapping the distribution of shaking and comparing it to the shaking patterns from modern earthquakes with known magnitudes.

To do this, researchers use descriptions of what happened to assign a number, called the intensity of shaking, at locations where there are accounts. Intensities are described by the Modified Mercalli scale (table 5.1), which uses

TABLE 5.1 The Modified Mercalli Intensity Scale

I. *Shaking not felt, no damage*: Not felt except by a very few under especially favorable conditions.

II. *Shaking weak, no damage*: Felt only by a few persons at rest, especially on upper floors of buildings.

III. Felt quite noticeably by persons indoors, especially on upper floors of buildings. Many people do not recognize it as an earthquake. Standing motorcars may rock slightly. Vibrations similar to the passing of a truck. Duration estimated.

IV. *Shaking light, no damage*: Felt indoors by many, outdoors by few during the day. At night, some awakened. Dishes, windows, doors disturbed; walls make cracking sound. Sensation like heavy truck striking building. Standing motorcars rocked noticeably.

V. *Shaking moderate, very light damage*: Felt by nearly everyone; many awakened. Some dishes, windows broken. Unstable objects overturned. Pendulum clocks may stop.

VI. *Shaking strong, light damage*: Felt by all, many frightened. Some heavy furniture moved; a few instances of fallen plaster. Damage slight.

VII. *Shaking very strong, moderate damage*: Damage negligible in buildings of good design and construction; slight to moderate in well-built ordinary structures; considerable damage in poorly built or badly designed structures; some chimneys broken.

VIII. *Shaking severe, moderate to heavy damage*: Damage slight in specially designed structures; considerable damage in ordinary substantial buildings, with partial collapse. Damage great in poorly built structures. Fall of chimneys, factory stacks, columns, monuments, walls. Heavy furniture overturned.

IX. *Shaking violent, heavy damage*: Damage considerable in specially designed structures; well-designed frame structures thrown out of plumb. Damage great in substantial buildings, with partial collapse. Buildings shifted off foundations.

X. *Shaking extreme, very heavy damage*: Some well-built wooden structures destroyed; most masonry and frame structures destroyed with foundations. Rails bent.

XI. Few, if any (masonry) structures remain standing. Bridges destroyed. Rails bent greatly.

XII. *Damage total*: Lines of sight and level are distorted. Objects thrown into the air.

roman numerals from I (generally unfelt) to XII (total destruction). As shown, significant shaking is required to cause damage. Thus, in the December earthquake, New Madrid itself experienced violent shaking (intensity IX) that caused heavy damage, whereas St. Louis, Louisville, and Nashville experienced strong shaking (intensity VI–VII) that caused light damage.

The intensity values are plotted on a map, and contour lines called isoseismals are drawn separating areas of different average intensities. Naturally, there's some uncertainty involved. Two people reading the same account can choose slightly different intensity values. Nearby places can have different intensities because of soil conditions. Where there aren't many reports, as occurred in 1811 and 1812 because much of the country was sparsely settled, a fallen chimney can raise the value for a large area. Chimneys are often the first parts of a building to be damaged because they're made of brick, which is easier to damage than wood, and because they're at the top of a building and so move the most when the ground shakes.

Hough's analysis gave a magnitude of about 7.2 for the December earthquake. Shaking from the January and February shocks is described as being less and more, respectively, than the December one. Using the same approach, Hough found magnitudes of 7.0 and 7.4 for the later earthquakes.

For simplicity, we'll use these values throughout this book. They're essentially the same as the ones that Otto Nuttli originally found for the three shocks: 7.2, 7.1, and 7. 4. However, they're much lower than the ones that Arch Johnston favored: 8.1, 7.8, and 8.0. The lower values make more sense to me for several reasons, including the fact that the intensity pattern looks somewhat like that of a magnitude 7.2 earthquake that happened off the east coast of Canada in 1929. The magnitudes make a big difference because the higher values were used in the hazard map showing that the New Madrid zone is more dangerous than California.

As a result, there will likely continue to be new estimates. As this book was in production, Sue Hough reported a new analysis that finds even lower values, about 6.8–7.0. This trend isn't unusual because the estimated magnitudes of historical earthquakes often decrease as more detailed studies are conducted. Estimating magnitudes is a complicated business, as we'll discuss in Chapter 8. Even with modern seismograms, there's an uncertainty of about 0.2. Using historical accounts, the uncertainty is bigger, at least 0.3. Different analyses estimate different values, which don't mean much unless they're larger than the uncertainty.

All this might sound "squishy." Because popular images of science come from television, where people in lab coats tend complicated machines, getting numbers from old letters and newspapers might not sound like "real" science. In fact, this is very real science. Even high-tech digital seismometers and super-precise GPS satellites can't determine the size of earthquakes that happened long before they were invented. As a result, intensity data give the best idea of what happened and are valuable in considering what to expect in possible future earthquakes.

Chapter 6

Breakthrough

It seems probable that a very long period will elapse before another important earthquake occurs along that part of the San Andreas rift which broke in 1906; for we have seen that the strains causing the slip were probably accumulating for 100 years.

—State Earthquake Investigation Commission report on the 1906 San Francisco earthquake

---

Life remained hard along the Mississippi even after the largest earthquake on February 7, 1812. Smaller earthquakes continued for months, as described by the famous naturalist John James Audubon, "Shock succeeded shock almost every day or night for several weeks, diminishing however, so gradually, as to dwindle away into mere vibrations of the earth. Strange to say, I for one became so accustomed to the feeling, as rather to enjoy the fears manifested by others."

The aftershocks left people fearful for the future. Some had lost homes and were living in tents. Some farms were significantly affected when land plunged into the river or was flooded. However, by summer life began to improve.

Setting a pattern for future disaster relief, the federal government eventually tried to help, but problems arose. In Myron Fuller's words, "The loss and suffering were eventually brought to the attention of Congress, but in the light of subsequent events it is not certain to what extent assistance was the real object of the agitation or to what extent it was a pretext for land grabbing on the part of certain unscrupulous persons."

A new law gave people whose lands had been destroyed certificates to replace them with other public land. However, as Fuller explained, because "people had become adjusted to the new conditions and the prospects for the future looked fairly bright," most of those eligible stayed and sold their certificates for a few cents per acre. Of 516 certificates issued, the original claimants used only 20. Speculators in St. Louis acquired most of the others, and "perjury and forgery became so common that for a time a New Madrid claim was regarded as a synonym for fraud."

Recovery from the earthquakes, as is usual after disasters, was fast. Humans are energetic and optimistic. They quickly get to work rebuilding their lives. This speed often surprises scientists who study disasters for many years to learn more about what happened and why. When new methods become available, researchers often go back and study past disasters. They're often studying a disaster long after those involved have almost forgotten the event or died.

The earthquakes of 1811–1812 happened as geology as a science was just beginning. Geology continued to develop over the century. Geologists learned to recognize rock layers, map their distribution, study their ages, and interpret how processes in the earth formed and changed them. Still, although people made excellent observations of what they saw and felt during and after earthquakes, what caused earthquakes remained a mystery.

This was true even for pioneering geologist Charles Lyell, who visited the New Madrid area in 1846. Lyell's classic book, *Principles of Geology*, inspired generations of successors including Charles Darwin. The book described earthquakes but couldn't explain what caused them. Lyell decided that some unknown process must be involved because "the chemical and mechanical changes in the subterranean regions must often be of a kind to which no counterpart can possibly be found in progress within the reach of our observation." He favored research because he realized that recognizing what isn't known is important for progress. Thus, "speculations on these subjects ought not to be discouraged, since a great step is gained if they render us more conscious of the extent of our inability to define the amount and kind of results to which ordinary subterranean operations are now giving rise."

Not understanding why earthquakes occurred was a huge limitation for the young science. Without an explanation, geology couldn't say anything useful about where and when earthquakes were likely.

Disaster

The science started coming together on April 18, 1906. At 5:12 AM, a violent earthquake shook San Francisco and the surrounding area. John Farish, a mining engineer who was staying at the St. Francis, one of the city's finest hotels reported:

> I was awakened by a loud rumbling noise which might be compared to the mixed sounds of a strong wind rushing through a forest and the breaking of waves against a cliff. In less time than it takes to tell, a concussion, similar to that caused by the nearby explosion of a huge blast, shook the building to its foundations and it began a series of the most lively motions imaginable. Accompanied by a creaking, grinding, rasping sound, it was followed by tremendous crashes as the cornices of adjoining buildings and chimneys tottered to the ground.

Buildings collapsed, killing people and trapping others. Witnesses' descriptions are vivid: "The whole street was undulating as though the waves of the ocean were coming toward me." "I saw the whole city enveloped in a pile of dust caused by falling buildings." Almost immediately, fires broke out as natural gas lines burst, electric lines fell, and cooking stoves toppled. Firemen connected hoses to hydrants, but no water came out because all but one of the water mains had broken. The fires burned for two more days until stopped by a combination of rain, firefighting, and running out of things to burn.

When it all was over, much of the city was in ruins. About 3,000 people are thought to have died, more than in all other U.S. earthquakes to date combined. At least half of the city's 400,000 people were homeless, and 28,000 buildings were destroyed. An estimated $500 million, $10 billion in today's dollars, in damage occurred.

The earthquake affected survivors in many ways. Four-year-old Ansel Adams crashed into a garden wall, leaving the famous nature photographer with an "earthquake nose." Five-year-old Cecil Green acquired an interest in geology, which led to a career that included developing geophysical instruments in his company that became Texas Instruments and becoming a major donor to earth science programs.

Elastic Rebound

The earthquake catalyzed the development of earthquake science. Although some people still blamed earthquakes on God's will, by 1906 most accepted them as a natural phenomenon. Noting the survival of a big liquor warehouse, local poet and jokester Charles Kellogg Field wrote:

If, as they say, God spanked the town
For being over frisky,
Why did He burn the churches down
And save Hotaling's whiskey?

Within five months, the spirit of scientific inquiry led to the formation of the Seismological Society of America, which remains active today. Dues were $2 a year, compared to $100 today. In addition to scientists and engineers, some early members were interested members of the public. One was John Muir, who founded the Sierra Club. Muir had experienced a major earthquake in 1872 in Yosemite valley, when he "ran out of my cabin, both glad and frightened, shouting, 'A noble earthquake! A noble earthquake!' feeling sure I was going to learn something."

Geologists were eager to investigate the earthquake. Within a few days, Andrew Lawson, head of the University of California geology department, persuaded the governor to establish a study commission. The governor set up the commission but provided no funds, which fortunately came from the Carnegie Institution of Washington. The commission's work set the standard for future earthquake studies because it combined two powerful methods.

One is field geology, detailed observations of the landscape and rocks. As Fuller's studies of the New Madrid area almost a hundred years after the 1811–1812 earthquakes show, geologists can learn a lot about what happened not only in a recent earthquake but also in earlier ones. Field geology is a special skill that most seismologists—myself included—don't have. We generally lack the observing powers. One of my fellow graduate students was having trouble finding the boundary between two rock units until the instructor pointed out, "if that contact was a rattlesnake, you'd be dead!"

The other method is geophysics, which uses data from seismometers and other observing systems to study the physics of earthquakes. Geophysicists typically get into the subject because they like math and physics but are fascinated by our planet. Mountains and faults seem more real to us than invisible subatomic particles. As another fellow student told me, "seismology is the next best thing to being a professional backpacker."

Both methods continue to yield important discoveries. Some come from new technology like digital seismometers, GPS, radioactive age dating, and satellite radar—and others come from the insightful eyes of field geologists. In geology, the whole is always greater than the sum of the parts.

The commission's leading field geologist was G. K. Gilbert from the University of California. (There's no need for "at Berkeley" because at the time there was only one campus.) Although he'd studied faults, he was disappointed not to have felt a major earthquake and so explained, "It is the natural and legitimate ambition of a properly constituted geologist to see a glacier, witness an eruption, and feel an earthquake . . . When, therefore I was awakened in Berkeley on the eighteenth of April last by a tumult of motions and noises, it was with unalloyed pleasure that I became aware that a vigorous earthquake was in progress."

Gilbert and colleagues immediately began studying the earth's effects and found a fascinating pattern. At many places, the ground was broken (fig. 6.1). Landmarks like fences, pipes, roads, or train tracks that used to be straight

FIGURE 6.1 *Left,* Ground breakage along the fault; *Right,* A fence offset about 11 feet by the earthquake. (Commission report)

had been offset, with their west sides moved north relative to their east sides by an average of about 13 feet. Even more amazing, these sites lined up for hundreds of miles along a zone that became known as the San Andreas fault (fig. 6.2).

The geophysical side of the investigation used results from geodesy, the science of the shape of the earth. Long before GPS, accurate measurements could be made with surveying equipment. Fortunately, the Coast and Geodetic Survey had surveyed the area twice before the earthquake. Resurveying the area after the earthquake showed that the motion across the fault extended more than 30 miles on either side of the fault. The offset fences were just the tip of the iceberg. Something even bigger had happened.

Commission member Harry Reid, a professor from Johns Hopkins University, used these observations to propose the theory that seismologists still use to explain earthquakes on a fault. In this theory, called elastic rebound, rocks on opposite sides of the fault move in opposite directions. This motion causes forces, called "stresses," on the fault. Stress is the force per unit area inside a solid material.

However, friction on the fault "locks" it, so the rocks right at the fault can't move. This deforms the rock in a wide zone on either side of the fault. This deformation, a change in shape, is called "strain." As time goes on, more strain accumulates. Eventually, the strain overcomes the friction and the fault slips, causing an earthquake.

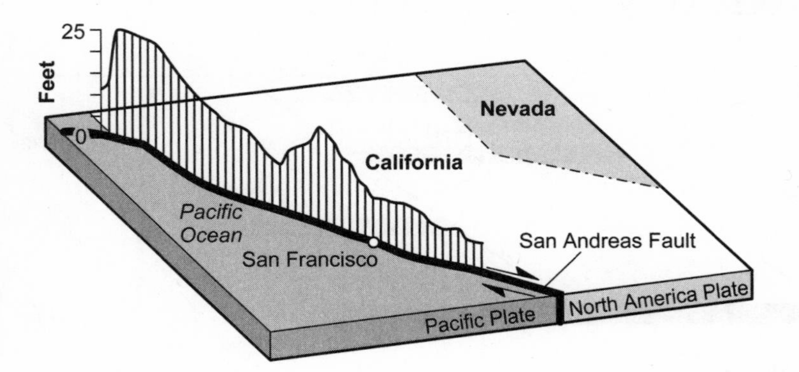

FIGURE 6.2 Horizontal slip along the San Andreas fault in the 1906 earthquake. (After USGS)

To see how this works, imagine a straight fence built across a fault (fig. 6.3). Over hundreds of years, the rocks on either side move, leaving the fence distorted near the locked fault. The strain extends for tens of miles on either side of the fault. Eventually, the earthquake happens, and the fault slips. The rocks on either side snap back, so the ground at the fault catches up in seconds to the rocks farther away, leaving the fence offset across the fault. This process, called the earthquake cycle, repeats as long as the rocks on either side of the fault continue sliding.

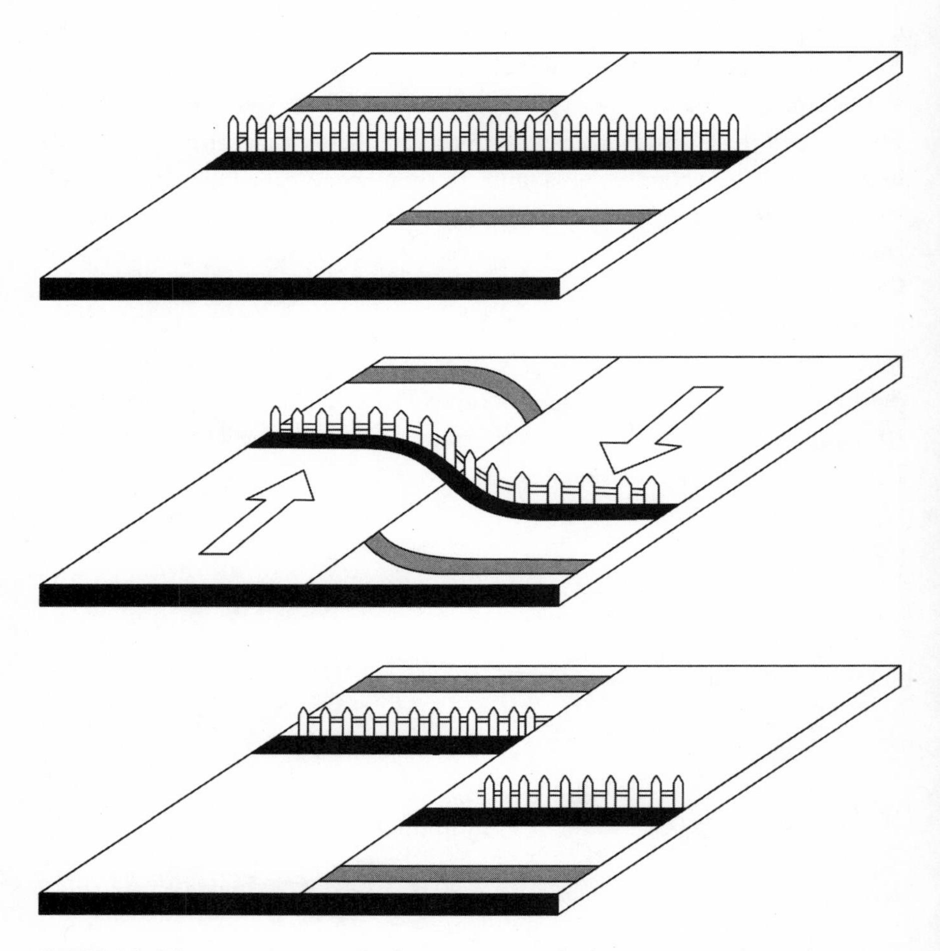

FIGURE 6.3 The movement of a fence across a fault over time shows how elastic rebound works. (Stein and Wysession, 2003)

This word "elastic" describes the way the ground distorts before the earthquake as though it were made of rubber. The distortion is much less, of course. Pulling at the ends of a rubber band stretches it a lot more than pulling on the ends of a piece of rock. Still, rocks do stretch and bend, as shown in figure 6.4. The word "rebound" describes the way the rocks on either side of the fault snap back during the earthquake. Another phrase for this behavior is "stick-slip": the fault is stuck for a long time and then slips.

I illustrate elastic rebound in class by attaching a rubber band to a soap bar in a box and pulling it across a yoga mat (fig. 6.5). Because of friction between the box and the mat, the box is stuck and doesn't move while the rubber band stretches. As the rubber band stretches, or strains, energy is stored in the rubber band. The more it stretches, the greater the force it applies to the box. Eventually, the force overcomes the friction. The rubber band snaps back to its original length, and the box suddenly slips forward. This analogy gets to the heart of how faults and earthquakes work.

Elastic rebound was the breakthrough seismology needed. It explained the mystery of how earthquakes happened. People had seen broken ground at the surface after earthquakes before, but thought that it was a side effect of

FIGURE 6.4 Folded rocks along the San Andreas fault near Palmdale, California.

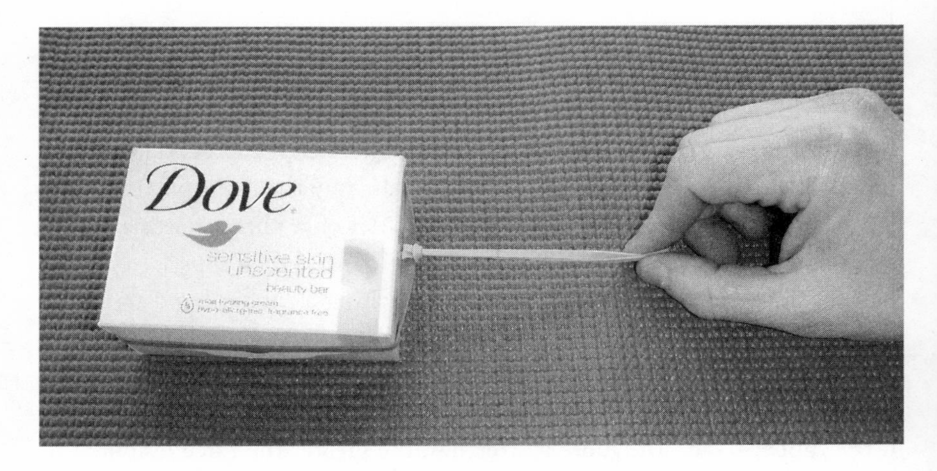

FIGURE 6.5 An analogy for elastic rebound.

an earthquake. Elastic rebound showed that the broken ground resulted from the fault motion that caused the earthquake.

The earthquake cycle is a great example of one of geology's most important concepts—slow processes over long times have big effects. The slow motion across the fault can't be seen without special equipment, but its effects are dramatic. The sides of the fault move very slowly—inches per year—but they add up to cause major earthquakes. In the same way, slow motions build mountains and create oceans.

Using Elastic Rebound

Suddenly, seismology could say many useful things about earthquakes. The commission had wanted to find "evidence upon which a judgment might be based as to the probability of recurrence of the earthquake in the future." Now, they had it.

Elastic rebound showed that strain on the fault built up slowly over many years. Because the fault had just slipped in the earthquake, it wouldn't slip again for a long time. Reid estimated that strain had been accumulating for about a hundred years before 1906, so a similar earthquake wouldn't happen soon. Fusakichi Omori, Japan's leading seismologist, who had joined the commission, agreed that the area "will be free of these great earthquakes for fifty years or more." They were right, since it's already been more than a hundred years since 1906.

If this insight had been available sooner, science could have shaped better public policy. San Francisco's city government and civic boosters stressed that most of the damage came from the fire rather than the earthquake. A spokesman for the powerful Southern Pacific Railroad said that it was important "to keep California and San Francisco from being misrepresented by the sensation mongers. We do not believe in advertising the earthquake. The real calamity was undoubtedly the fire." The city feared that if businesses thought another earthquake would happen soon, they wouldn't invest there. In particular, investors wouldn't buy bonds that the city was selling to pay for rebuilding. Fire wasn't as much of a problem because fires also occurred in other cities.

What the city said was true but misleading because the earthquake both started the fires and allowed them to spread. Once seismologists understood elastic rebound, there was no need to downplay the earthquake. Investors could have bought bonds knowing that no similar earthquake would happen on a time scale that mattered to them.

Understanding elastic rebound also let the commission give calm advice about rebuilding the city. Buildings needed to be built carefully, but there was no need to rush because it would be a long time before another large earthquake. Because buildings on weak soil or landfill had suffered much greater damage than ones on solid rock, they recommended "studying carefully the site of proposed costly public buildings where large numbers of people are likely to be congregated. In so far as possible such sites should be selected on slopes upon which sound rock foundation can be reached. It is probably in large measure due to the fact of their having such a rock foundation that the buildings of the State University, at Berkeley, escaped practically uninjured."

It was also important to build better because "a great many of our schools proved to be of flimsy construction and ill adapted to meet the emergency of an earthquake shock of even less severity than that of the 18th of April."

Most important, elastic rebound gave a method to tell whether and when another big earthquake would happen. The 13 feet that the fault had moved in a few seconds had accumulated over many years. That meant that measuring the slow motion across the fault could tell how long it would be until another earthquake that big could happen. It's like saving up to buy something:

from the price and how much you're setting aside every month, you know when you can afford it.

Using this idea, the commission recommended that geodetic measurements be made to measure the motion across the fault over time as strain built up before a future earthquake. They suggested that "we should build a line of piers, say a kilometer apart" across the fault so "careful determination from time to time of the directions of the lines joining successive piers . . . would reveal any strains which might be developing." That's done today, using GPS as a high-tech substitute for piers. GPS measurements show movement across a zone of about 30 miles on either side of the San Andreas fault as strain builds, indicating that a major earthquake is on the way.

GPS also shows strain accumulating across other faults on which major earthquakes are expected. For example, even before the San Francisco earthquake, Gilbert had the idea that large earthquakes in Utah happened on faults that raised the Wasatch Mountains over the Salt Lake City valley. People living there were skeptical because they felt at most small earthquakes. However, GPS shows the motion that will give rise to future earthquakes.

This is why my friends and I were so surprised to find no motion in the New Madrid seismic zone. Because there had been large earthquakes in the past, we had expected to find strain accumulating. The fact that there's no motion shows that strain isn't building up, so there's no reason to expect an earthquake soon.

The Missing Piece

Even today, geologists are impressed by the brilliant job the commission did. Their results started modern earthquake science and remain crucial to everything we do. Several members were honored in different ways. The relation that describes how the rate of aftershocks slows down after a large earthquake is called Omori's law. The Seismological Society of America awards the Reid medal for outstanding contributions in seismology and earthquake engineering. The Geological Society of America gives the Gilbert award for discoveries in planetary geology because of another of Gilbert's accomplishments: recognizing that impacting objects caused the craters on the moon. Andrew Lawson might be the most honored; rumor says that he named the

San Andreas fault after himself, rather than the San Andreas valley through which it runs. True or not, the story has the same effect.

The commission got only one major thing wrong, and no one blames them. Because the geodetic data only extended about 30 miles on either side of the fault, they "throw no light on what occurred at greater distances . . . There is no reason to believe that other ruptures or slip occurred outside the region." Not knowing any better, the commission assumed that the motion occurred only in a narrow zone across the fault.

That left a big piece of the earthquake puzzle missing: What makes the fault move? The answer was so outrageous that the commission never thought of it. They were thinking too small. It took 60 years for geologists to figure out that motion on either side went huge distances. We now know that the entire Pacific is slipping north relative to all of North America. This motion between two great plates explains why the two sides of the fault moved.

Chapter 7

How the Ground Shakes

> The earth came rolling towards us, like a wave on the ocean.
>
> —John Walker's description of the December 16, 1811, New Madrid earthquake

Accounts of the ground shaking in the New Madrid and San Francisco earthquakes seem almost unbelievable. To make sense of them, we need to ask three questions. First, how did solid ground shake enough to do so much damage? Second, what happened on the fault during an earthquake? Third, what made the fault move?

Answering these questions lets us understand what happened in the earthquakes of 1811–1812 and what could happen if such an earthquake happened again. The next few chapters explore these questions, starting with how the ground shakes.

Seismic Waves

Ground shaking happens when the energy that was stored up over many years on a locked fault is released in an earthquake. Some of the energy goes into moving the two sides of the fault, and some travels away from the fault into the surrounding rocks. The traveling energy is carried by seismic waves. Although these sound mysterious, they're easy to understand using some familiar ideas.

The best place to start is with a Slinky, the toy spring that seismologists love because it shows beautifully how seismic waves work. It's hard to imagine teaching about waves without one. I use two, one with the logo of the Incorporated Research Institutions for Seismology on it. Slinkys are easy to find because more than 250 million have been sold since they were invented in 1943, including ones made of 14-karat gold. A Slinky is shown on a U.S. postage stamp, and one has flown on a space shuttle.

Imagine laying a Slinky on a table and pulling on one end (fig. 7.1). The pull makes a pulse move along the spring. If you look carefully, you see the pulse is a part of the spring that is extended because nearby coils move apart. The extended coils then pull back together, which pulls the next group of coils apart. This makes the pulse of extension move along the Slinky. Each part of the Slinky moves only a little back and forth, but the pulse moves from one end to the other.

Something similar happens if you shake one end of a stretched string. The shake causes a pulse that moves along the string. Each part of the string moves

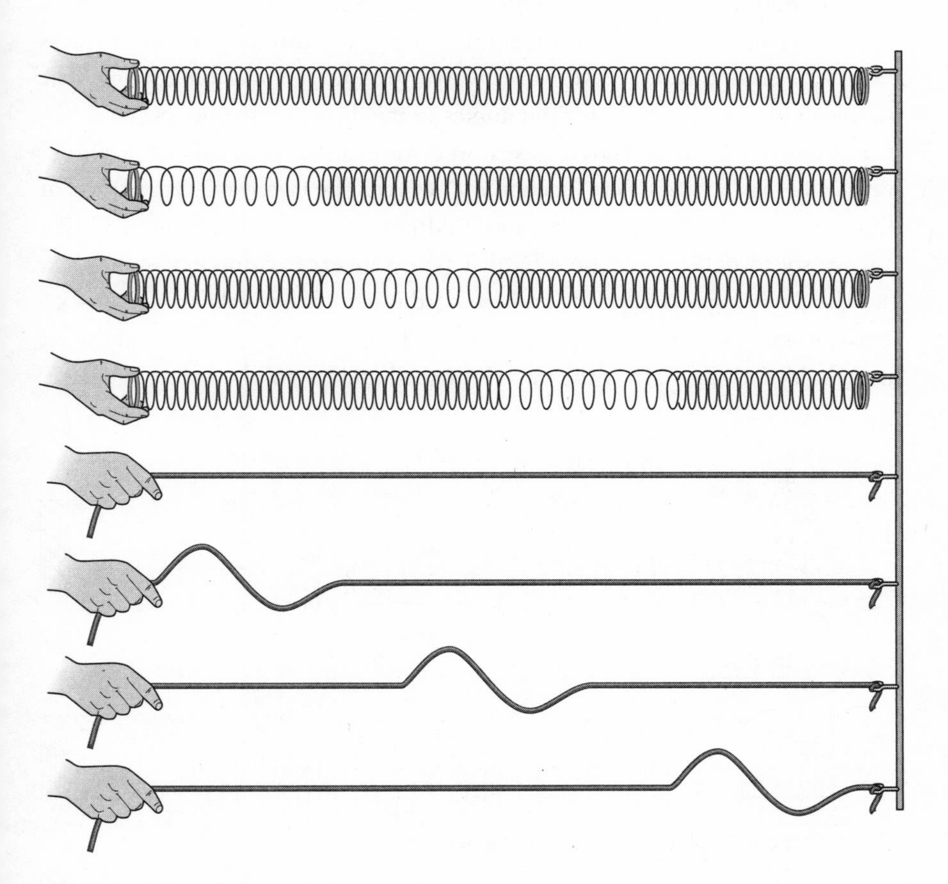

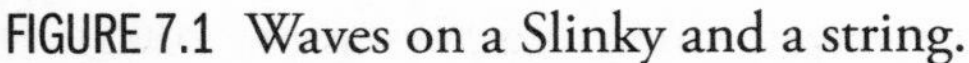

FIGURE 7.1 Waves on a Slinky and a string.

up and down—not sideways—but the pulse moves sideways along the string.

Although the Slinky and the string are different, there's a pattern here. A pulse, or disturbance, travels all the way through, even though any particular part doesn't move that far. The pulses are called waves.

An important feature of waves is that their speeds depend on the properties of the material they travel through. The wave speed on a string depends on how tightly the string is stretched: the tighter the string, the faster the wave. Similarly on a Slinky, the wave speed depends on the strength of the spring: the stronger the spring, the faster the wave.

The Slinky and string show the two kinds of seismic waves that travel through solid rock after an earthquake (fig. 7.2). The first waves, known as "P" or compressional waves, are like the ones on a Slinky. As these waves travel, material in the rock expands or contracts along the direction that the waves move. The second kind, "S" or shear waves, are like the ones on a string. The material in the rock moves at right angles to the direction the waves travel.

P waves travel faster than S waves, so P waves used to be called "primary" waves whereas S waves were "secondary" waves. The two wave types travel at different speeds because they depend on different properties of the rock. The P wave speed depends on how hard it is to squeeze—compress—the rock, and the S wave speed depends on how hard it is to bend—shear—the rock.

To understand how waves work, think about a spring with a weight, or mass, at its end (fig. 7.3). If the mass is pushed and let go, the spring oscillates

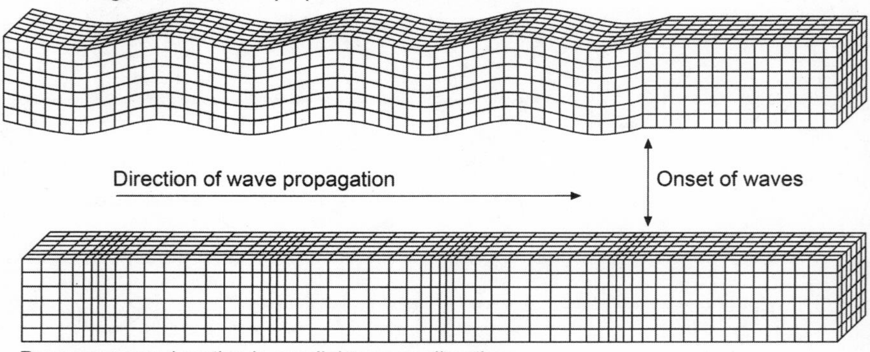

FIGURE 7.2 Compressional and shear waves. (Stein and Wysession, 2003)

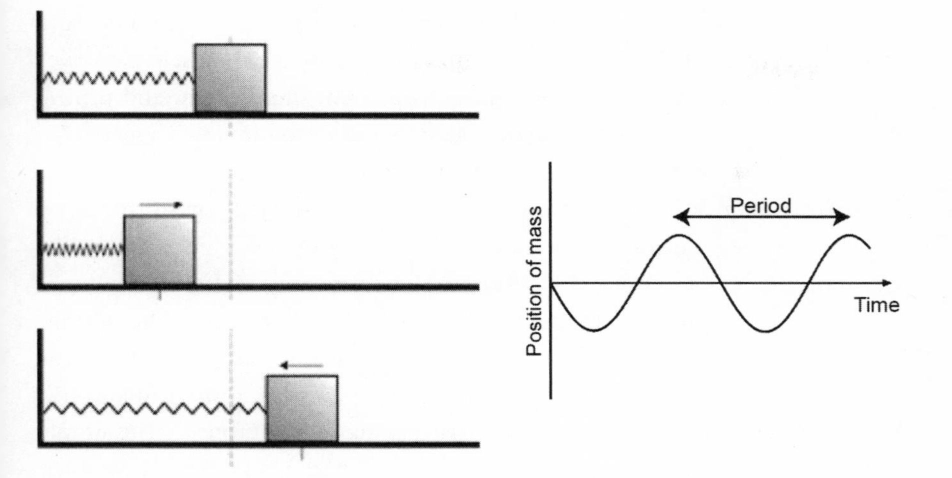

FIGURE 7.3 An oscillating spring and mass.

back and forth around the spring's unstretched length. The spring is first compressed and then pushes the mass back until it overshoots the unstretched length. At this point, the spring is extended, so it pulls the mass the other way. Thus, the spring exerts a restoring force that pushes or pulls the mass back to the original unstretched position.

Looking at how the mass moves over time shows the oscillation. Each back-and-forth cycle takes a length of time called the natural period. The period equals the square root of the mass divided by the strength of the spring. A stronger spring makes the system oscillate faster, with a shorter period. A heavier mass makes it oscillate more slowly, with a longer period. The spring and mass form what's called a harmonic oscillator, where "harmonic," as in "harmony," means that the oscillator has a characteristic period.

This motion is like waves on a Slinky or string, because waves involve harmonic oscillations. As a wave goes by, points on the ground move back and forth with a characteristic period. Waves are described by their periods, like a 20-second seismic wave.

Seismometers

How do we know all this? It's easy to see waves on a Slinky or string, but surely we can't see them in the earth. In fact, we can. That's what a seismometer does.

A seismometer is a sensor that responds to ground motion by producing a signal that's recorded along with the time, so we know when seismic waves arrived. The combination of the seismometer, recording system, and timing mechanism is called a seismograph. A seismograph produces seismograms—the wiggly lines seen in earthquake movies.

Measuring ground motion is hard. We measure the motion of things by comparing their position over time to something else. For example, without looking at objects out the window, there's no way to know if a car is moving because the speedometer and odometer only measure the wheels' turning. They would look the same if the engine was running but the car was up on blocks, going nowhere. The problem is that an instrument built to record the motion of the ground is also sitting on the ground. The simple but brilliant solution that makes a seismometer work is to use a spring-mass system.

The guts of a seismometer (fig. 7.4) are a spring attached to an outer frame, with a mass attached to the spring. The mass's position over time is recorded by a system like the rotating drum shown. When the ground isn't shaking, the frame doesn't move, so the mass doesn't move, and a flat line is recorded. When the ground shakes, a wiggling line is recorded.

The spring is what makes this work. If the spring were a rigid metal rod, the mass would move just the way the ground does, so the seismometer wouldn't record anything. However, because the spring stretches, the mass moves in a different way from the ground.

To see this, hang a coffee cup from a strong rubber band that you're holding in one hand. With the other hand, quickly pull the cup down and let go.

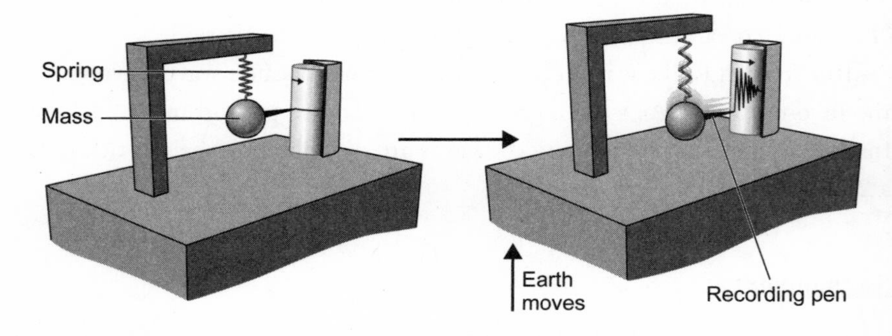

FIGURE 7.4 How a seismometer works.

The cup will oscillate up and down. The time for each oscillation is the natural period. Now, if you move your top hand up and down, the cup oscillates with the period that your hand moves. What's neat is that how much the cup moves depends on how fast you move your hand. It moves most when your hand moves at the natural period. What you're doing is like "pumping" a backyard swing by pushing each time the swing returns to its highest point. Pushing a spring-mass system at its natural period causes the largest motion. This idea is called resonance. Pushing this system at other periods causes less motion.

All of this is described by math called differential equations that college students who want to become seismologists learn. These equations tell how any spring-mass system responds when a force is applied. They're used to design seismometers that respond to waves of certain periods as well as to figure out from seismograms how the ground actually moved. Earthquake engineers use them another way. They describe buildings as spring-mass systems and use the math to figure out how much a building will shake if earthquake waves of different periods hit it.

A challenge in building seismometers is that the ground motions from distant earthquakes are typically very small, thousandths of an inch or less. Seismographs magnify the ground motion so it can be seen and studied. This problem had been solved before the 1906 San Francisco earthquake, so the commission studying the earthquake had almost 100 seismograms from around the world. The early seismometers were mechanical systems that responded best to very large earthquakes because they could only magnify the ground motion about 100 times. Over the years, seismographs have gotten much better. Sensors have become much more sensitive, their output is recorded digitally, and signals from GPS satellites give very accurate timing.

Seismometers are showing up all over because many laptop computers have a seismometer inside. It's there to tell if the computer has been dropped and lock the hard drive before the computer hits the ground. A program called SeisMac, which is available free on the Internet, shows the motion on the screen. It and similar programs for PCs are becoming popular, and a project called the Quake Catcher Network lets people join their computers into a network that monitors earthquakes. There's also a seismometer application for iPhones, which uses the sensor that tells the phone which way the screen is oriented.

Seismograms

Seismograms are the language of earthquakes. Although at first they just look like wiggly lines, they're full of information. Figure 7.5 shows seismograms from a magnitude 5.2 earthquake in 2008 on the Wabash Valley fault system in southern Illinois. The earthquake shook buildings in Chicago and was felt as far away as Kansas and West Virginia but caused only minor damage near the epicenter. This earthquake was the largest in Illinois since the 1968 magnitude 5.4 one discussed in Chapter 4. Because it was typical of Midwest earthquakes, residents of the area where it happened took it in stride. One said, "It's not too much to get excited about. The ground's shook before, and it'll shake again."

These seismograms (fig. 7.5) were recorded 895 miles away in New Mexico. They show the ground motion in all three directions: east and west, north and south, and up and down. The first two are horizontal motions, along the earth's surface, and the third is vertical motion. The directions are called "components," so these seismograms are the east, north, and vertical (up) components.

The time given along the bottom is expressed as Universal Time, the technical name for Greenwich Mean Time. Universal Time avoids confusion with local time. This earthquake happened at 9:37 Universal Time near the border between Illinois, which is on central time, and Indiana, which is on eastern time. The bottom seismogram, the up component, shows that the P wave arrived in New Mexico a little after 9:40. The S wave arrived about three minutes later, and surface waves, mixtures of P and S waves, arrived at about 9:44.

P and S waves travel down into the earth and then back up, so they arrive at the seismometer from below. Because P waves involve motion in the direction that the wave moves, they show up best on the vertical component. Similarly, because S waves involve motion at right angles to the direction they travel, they show up best on the two horizontal components. If deciding which wave appears on which component seems confusing, you're in good company. Early seismologists didn't realize that seismic waves came up from below and so had trouble sorting this out.

The seismograms show that earthquakes generate S waves that are a lot bigger than the P waves. That's why people close to an earthquake feel the S wave much more than the P wave.

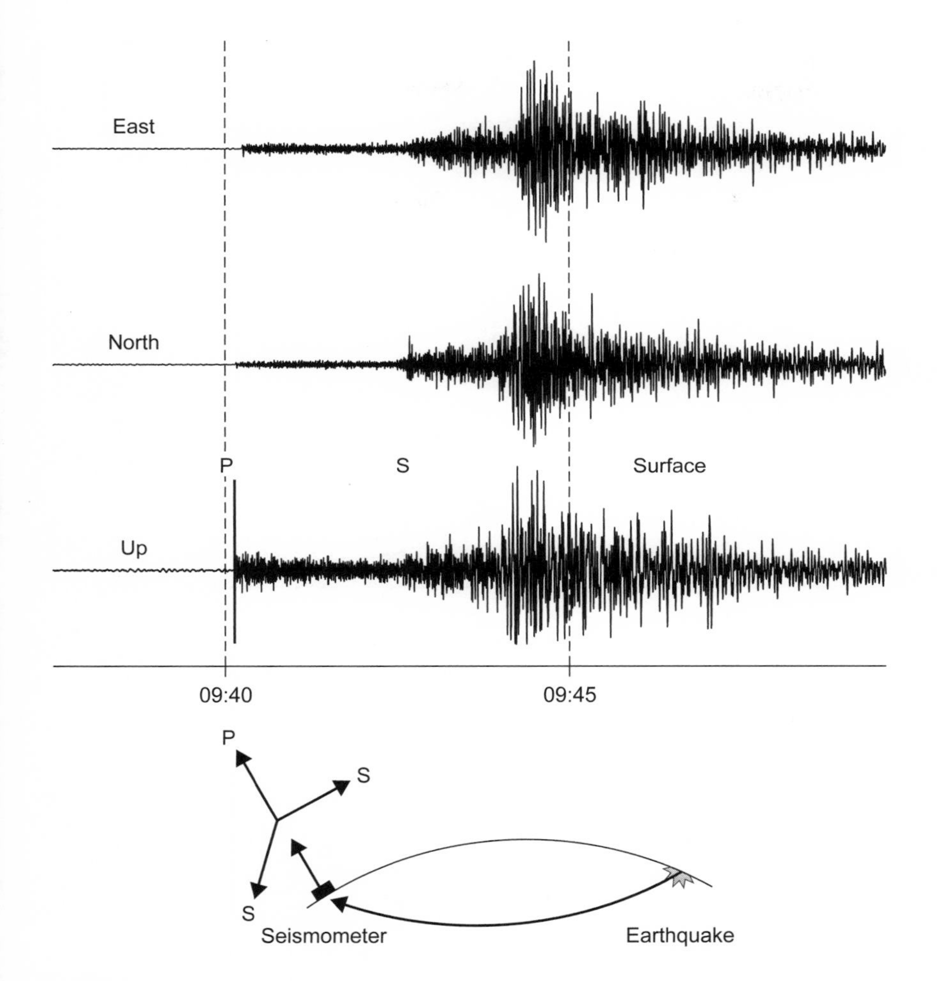

FIGURE 7.5 *Top,* Seismograms from an earthquake in southern Illinois on April 18, 2008, recorded in New Mexico (Figure by Carl Ebeling, data from Incorporated Research Institutions for Seismology); *Bottom,* Geometry of the waves' path, showing which waves appear on the vertical and horizontal components.

The circles on the map in figure 7.6 show how long the P waves took to reach different distances. The waves spread out from the earthquake like water waves in a pond after a rock is dropped.

As the map shows, seismic waves travel very fast. In about three minutes, P waves got about 900 miles from the earthquake, which is 18,000 miles per

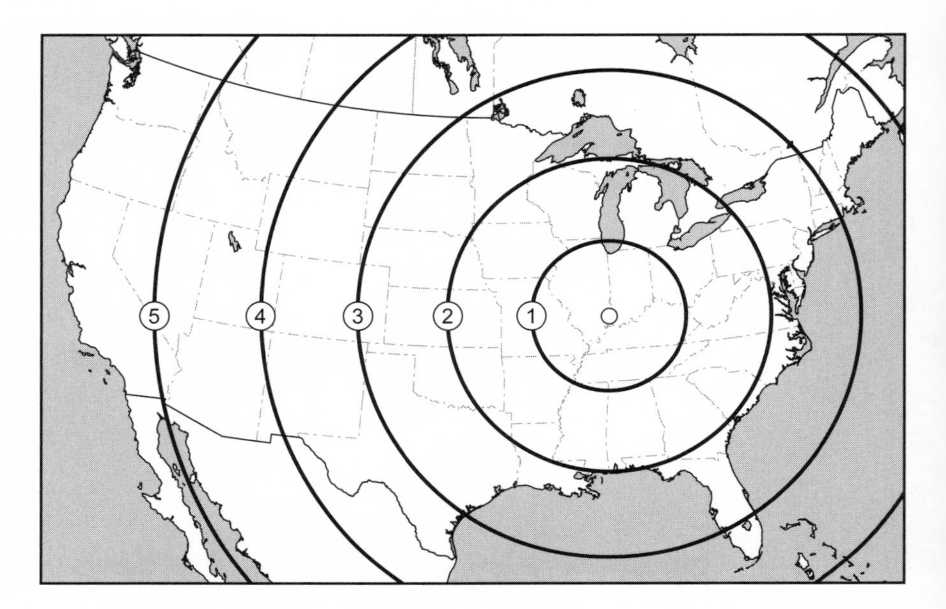

FIGURE 7.6 Circles showing the time in minutes after the earthquake when the first P wave arrived. (After USGS)

hour. They actually went even faster because they traveled a longer path down into the earth and back. Measuring many seismograms shows that the deeper seismic waves go in the earth, the faster they travel because the rocks are stronger at depth. Near the earth's surface, P waves move at about 3.4 miles per second, which works out to more than 12,000 miles per hour. S waves travel at about 2 miles per second, or more than 7,000 miles per hour. At the center of the earth, P waves travel 25,000 miles per hour. For comparison, jet airliners fly about 550 miles per hour, and the fastest airplane ever, the X-15 rocket plane, flew about 4,500 miles per hour.

The seismograms also show what an impressive piece of equipment a seismometer is. The ground motion was about 1/1000 of an inch, but the seismometer detected and recorded it.

Something that really helps us understand seismograms is that scientists have been studying waves for a long time. That's because waves are all around us. Even better, different kinds of waves behave in similar ways. That lets us use observations of familiar things to understand seismic waves.

The most familiar waves are water waves. If a rock is dropped into water, it generates a wave that moves outward. The water at any point moves up and down, but the wave travels a long way. Sound waves travel through air just like seismic waves in rock. Although the molecules in the air are too small for us to see them moving together and apart, we hear the pulse that results. We see due to light waves and talk on cell phones using radio waves.

The fact that different kinds of waves travel at very different speeds can be put to good use. A tsunami, the water wave generated by an underwater earthquake, travels at about 550 miles per hour, the speed of a jet airliner. Because seismic waves travel much faster, seismologists know that a large earthquake happened soon enough to warn places far from the earthquake that a tsunami might hit. For example, a P wave travels from Alaska to Hawaii in about 7 minutes, but a tsunami takes 5 hours. A tsunami warning system for the Pacific is based in Hawaii. This appeared in an episode of the TV show *Hawaii Five-0* in which a group of geology students called the "Brain Trust" plan to take over the warning center, issue a fake warning, and rob a jewelry store of $6 million in gems while people in Honolulu are fleeing to higher ground. Fortunately, the real warning center staff members are honest, though I might be biased because some took my classes years ago and one of my colleagues developed some of their software.

Another use of different waves' speeds comes from the fact that radio and other electromagnetic waves travel at the speed of light, an incredible 186,000 miles per second. Because this is much faster than seismic waves, automatic systems are being developed to locate an earthquake, decide if it's dangerous, and send a warning that would arrive tens of seconds before the seismic waves. Seismologists, engineers, and authorities are exploring what might be done with such warning times. Real-time warnings could be used to stop trains and shut down machinery like elevators or gas lines. The questions are whether the improved safety justifies the cost and whether the risk of false alarms is serious.

Why Places Shake Differently

We've talked about the fact that in the New Madrid earthquakes, the ground shaking was generally strongest close to the fault that moved and weaker at places farther away.

To see why, think of the wave produced by dropping a rock in water. As the wave spreads out in a circle, its height gets smaller. That's because of one of the great principles of science, conservation of energy, which says that energy can't be destroyed. Energy changes from one form to another but never vanishes.

To understand this, it's important to know what energy is. Although people talk about energy and have a vague idea what it is, it's hard to define. A good way to think about it is: Anything that can make something move involves energy. We store potential energy in the rock by picking it up. As it falls, the potential energy turns into kinetic energy, which is the energy of a moving object. When the rock hits the water, its kinetic energy provides the kinetic energy of the moving wave. As the wave spreads out in a growing circle, its total energy is fixed. Because the circle's circumference grows, there's less energy in each part, so the height of the wave gets smaller. In addition, some of the wave's kinetic energy is converted to heat energy by friction within the water. This is like the way rubbing your hands warms them. Eventually, the wave gets too small to see.

In the same way, seismic waves get smaller as they move away from earthquakes (fig. 7.7). That's why the 1811–1812 earthquakes caused severe shaking and heavy damage in New Madrid, whereas farther away places like St. Louis or Nashville experienced much less shaking and only minor damage. This is important when we think about damage from earthquakes, because a moderate sized earthquake nearby can do more damage than a larger one further away.

An interesting aspect of the way waves get smaller with distance is that shorter period waves decrease faster than longer period ones. We'll see in Chapter 15 that this is important because lower buildings are most vulnerable to damage by shorter period waves while tall buildings are more vulnerable to longer period waves.

Rocks in the central U.S. transmit seismic energy better than ones in the west. That's because friction in the colder rock of the Midwest, which is geologically less active, is less than in warmer rock from the more active west. To see this, look at figure 7.8, which shows seismograms from an earthquake in Texas. The seismogram from Missouri shows much more short-period shaking than the one from Nevada.

This difference is important in planning for earthquakes because it means that an earthquake in the Midwest will cause more shaking than one of the

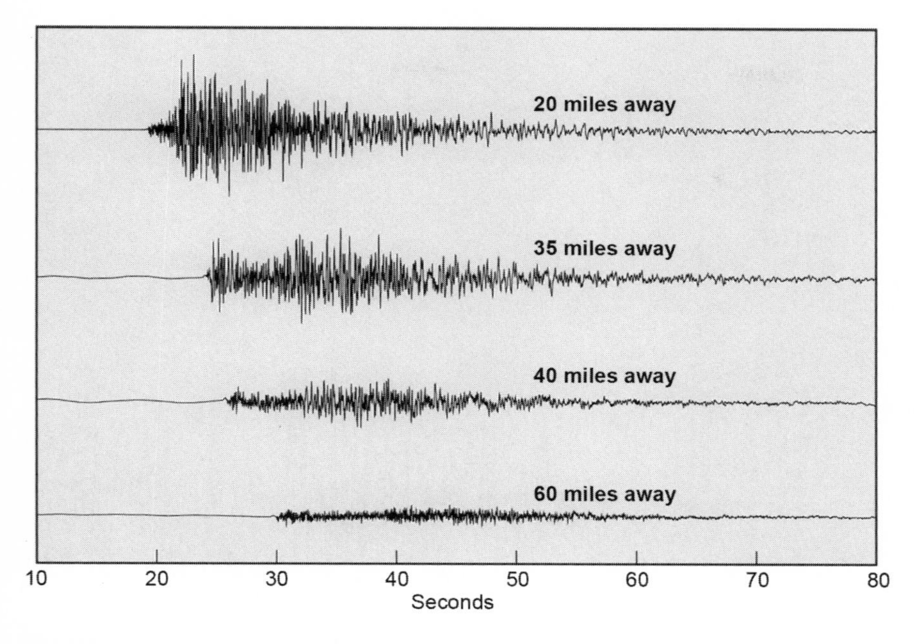

FIGURE 7.7 Seismograms showing how seismic wave amplitudes decrease at larger distances from an earthquake. (University of Nevada, Reno)

same size in California. Figure 7.9 shows this by comparing intensity maps for the December 16, 1811, New Madrid earthquake to the similar-size 1992 Landers earthquake east of Los Angeles.

Understanding waves also explains why in the 1906 San Francisco earthquake, buildings on landfill suffered more damage than ones on solid rock. We've all seen waves get taller and break as they approach a beach. That happens because the speed of a wave depends on the water depth, so a wave slows as it reaches the beach. The back part of the wave catches up with the front, so the wave gets taller (fig. 7.10). It's like the traffic jam that develops when cars exit from a fast highway to a slow street. That's why tsunamis are small and don't bother boats in the open ocean but get higher and destructive when they reach the shore.

Seismic waves behave the same way. Like waves on a string, seismic waves are faster in strong material like solid rock and slower in mud, landfill, or other soft material. Because they travel more slowly, they get bigger.

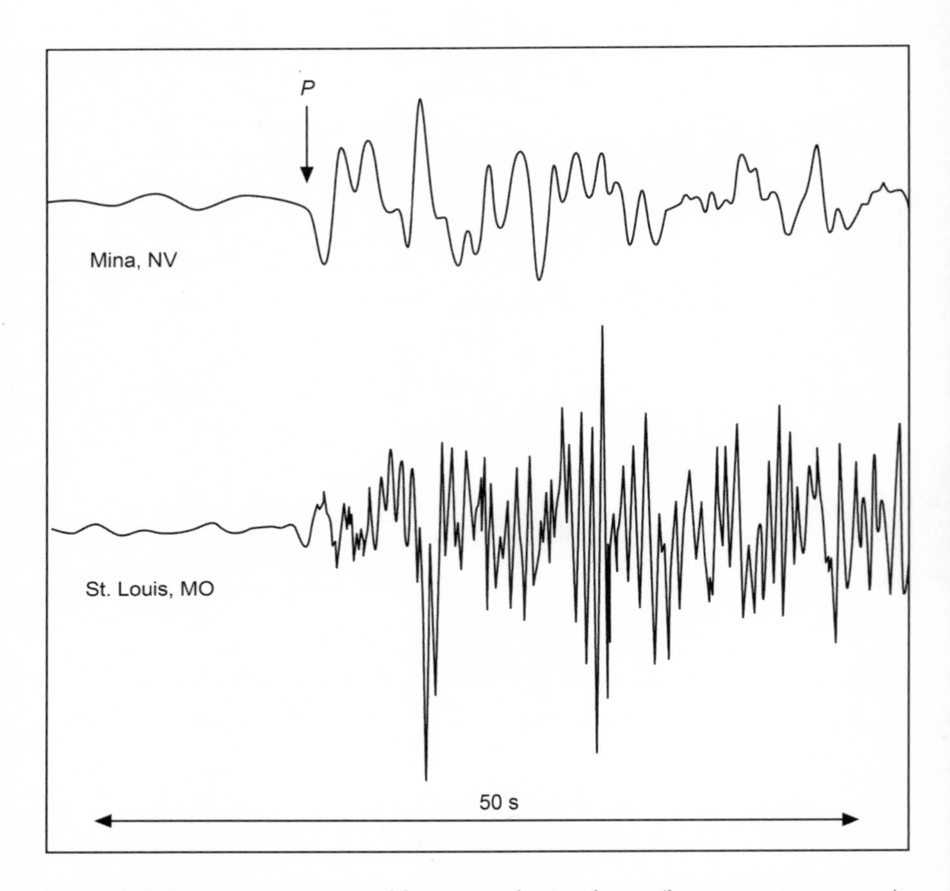

FIGURE 7.8 Seismic waves travel better in the Midwest (bottom seismogram) than in the west (top seismogram). (Stein and Wysession, 2003)

To see this, look at the seismograms in figure 7.10 from an aftershock of the 1989 Loma Prieta earthquake near San Francisco. Most of the deaths in the earthquake happened when a double-decked section of a highway built on weak dried mud collapsed. Nearby parts of the highway built on stronger material were damaged but didn't collapse. The seismograms show why; shaking is much greater on mud. For the same reason, building damage on filled land in the Marina district of San Francisco, where some of the fill is rubble from buildings destroyed in the 1906 earthquake, was greater than on nearby bedrock.

The fact that earthquakes produce stronger shaking on soft sediment is important in the New Madrid seismic zone because many people live along

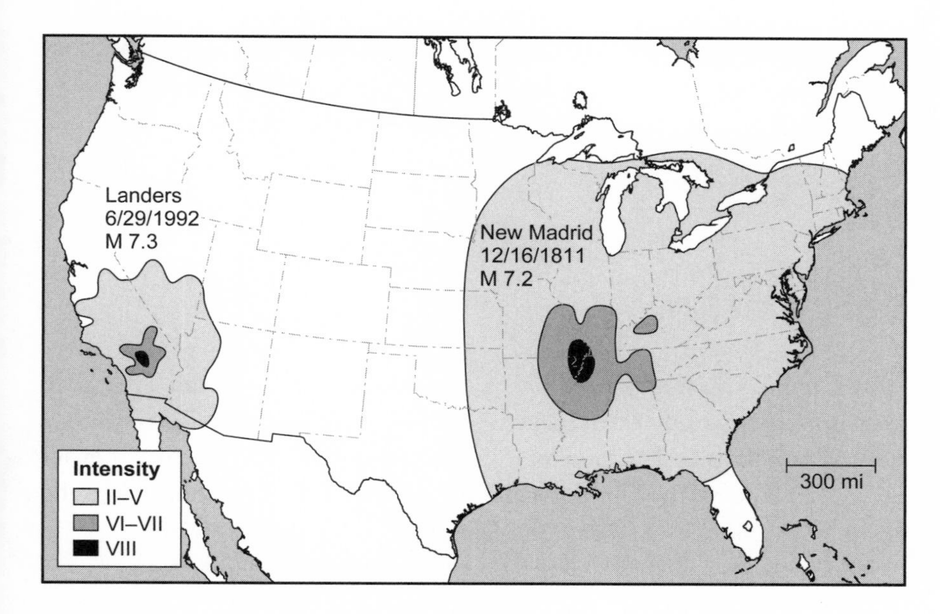

FIGURE 7.9 The December 1811 New Madrid earthquake shook a much larger area than the 1992 Landers earthquake. (Susan Hough)

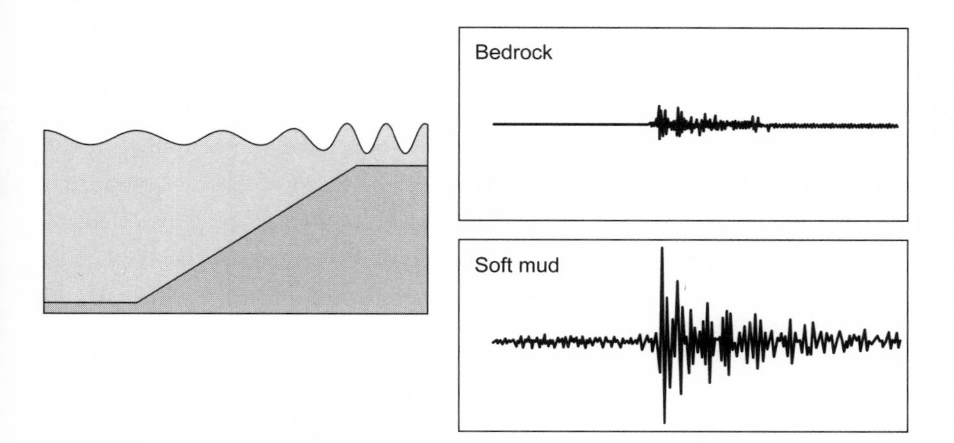

FIGURE 7.10 Just as water waves grow taller as they go from deep to shallow water (*left*), seismic waves get bigger when they travel from solid rock to weaker material. This effect explains why shaking on soft mud in the Loma Prieta earthquake was greater than shaking on solid rock, as shown on the right. (USGS)

the Mississippi and other river valleys. This effect has to be taken into account while evaluating reports of the shaking in 1811 and 1812 because the valleys were settled first. It's part of the reason some older studies found magnitude values that now seem far too high. It's also considered in designing buildings today.

Seeing into the Earth

At this point, you might be thinking that seismic waves are bad. True, they let seismologists study earthquakes. Otherwise, it sounds like all they do is destroy buildings and harm people. In fact, they're helpful because they let us see into the earth. That's great for science and does more to lower the prices of gasoline, natural gas, and heating oil than any politician.

The reason is that waves change when they go from one material to another in which they travel at a different speed (fig. 7.11). Part of the incoming wave is reflected back into the first material, and part is transmitted into the second. Both the angle and the size of the wave change, depending on the materials. When a wave enters faster material, the transmitted wave bends toward the boundary between the materials. This is why a pencil in water looks bent; light waves travel faster in air than in water. In the same way, lenses in eyeglasses or binoculars bend light to let us see better.

The size of the reflected and transmitted waves depends on how different the materials are in density and speed. If they're very similar, most of the wave is transmitted and only a little is reflected. This makes sense because if the two materials were the same, all of the wave would be transmitted, so nothing would be reflected. On the other hand, if the materials were very different, a lot of the incoming wave would be reflected and only a little would be transmitted.

These changes let us use seismic waves to map the earth's interior. As waves go down into the earth, they encounter rocks with different wave speeds and so change direction and size. Waves that took different paths through the earth reach seismometers at different times, so seismologists can tell which way each went.

The times when waves arrive and their sizes tell us about the seismic velocities at depth. The seismic velocities are used to locate earthquakes, as we'll discuss in the next chapter. Comparing the velocities to lab experi-

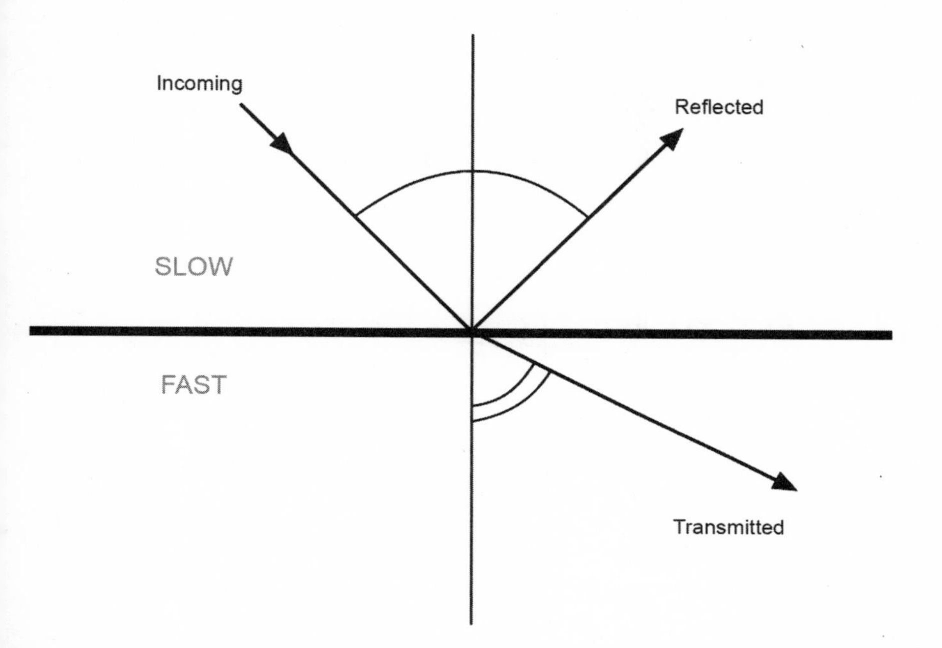

FIGURE 7.11 As a seismic wave goes from one material to another, part of the wave is reflected and part is transmitted. (Stein and Wysession, 2003)

ments tells us about the types of rock and the temperatures at depth. Because P and S waves travel at different speeds, combining them shows even more than each alone. In real life we can't go deep into the earth, despite the movie "The Core" that's often shown in geology departments' bad movie festivals. Instead, we get "pictures" of what is inside. This is a lot like the way doctors "see" inside people's bodies using electromagnetic waves (X-rays) and sound waves (ultrasound).

The basic picture is that the earth is made up of a series of layers, or shells, of different material (fig. 7.12). The outer shell is the crust, which is about 5 miles thick under the oceans and about 25 miles thick under the continents. Below that are the mantle, the liquid outer core, and the solid inner core. The crust and mantle are made up of different rocks, but the core is mostly iron. This picture tells a lot about how the earth works today—including part of the reason why earthquakes happen—and how it has evolved since its formation 4.6 billion years ago.

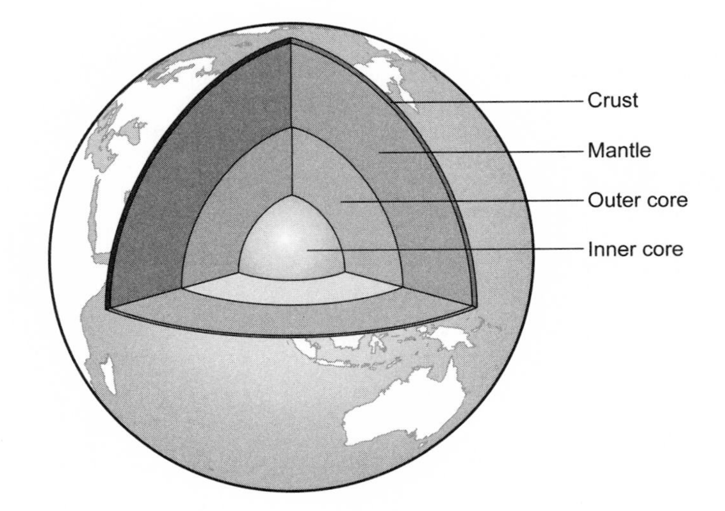

FIGURE 7.12 Schematic picture of the earth's interior based on data from seismic waves.

We'll see in Chapter 11 that a similar method lets seismologists see through thick Mississippi River sediment to study the buried faults that cause the New Madrid earthquakes. Because the faults are only a few miles below the earth's surface, seismologists don't need to wait for earthquakes that generate waves that go deep in the earth. Instead, seismic waves are generated using special equipment and recorded once they come back up after bouncing off different rock layers. The resulting pictures show how the faults moved over time. Oil companies also use this method to predict where drilling is likely to find oil or natural gas. Without it, finding oil would be much harder and more expensive.

Chapter 8

How Earthquakes Work

The joy of being a seismologist comes to you when you find something new about the earth's interior from the observation of seismic waves obtained on the surface, and realize that you did it without penetrating the earth or touching or examining it directly.

—Keiiti Aki (1930–2005), who first measured an earthquake's seismic moment

We've talked about seismic waves and how they shake the ground in earthquakes. The next question is how earthquakes generate these waves. Knowing this lets us figure out what could have happened in the 1811–1812 New Madrid earthquakes. That in turn lets us explore what caused them, whether and when they might happen again, and how dangerous that might be.

An earthquake is a complicated process, so describing it involves a number of questions. Let's start with the basic ones seismologists are used to being asked. Where was it? When did it happen? How big was it? What fault was it on? How did the fault move?

Answering these questions after an earthquake gives seismologists a chance to make fools of themselves on TV news. The first answers based on incomplete data sometimes aren't right. Eventually, better answers are determined by combining what's learned from the seismic waves that the earthquake generated with what's known about the geology of the area where the earthquake happened. That's tough for the 1811–1812 New Madrid quakes because they happened before the seismometer was invented, and there's a lot that isn't known about the area's geology.

In an earthquake, rocks on one side of a fault move relative to those on the other side, releasing stored energy that travels away as seismic waves (fig. 8.1). Studying these waves shows where the fault was, how it was oriented, and what motion occurred. The first step is locating the earthquake's focus or hypocenter, the place on the fault that first started moving. Maps often show the epicenter, which is the point on the earth's surface directly above the hypocenter.

A large part of the fault moves in a big earthquake, like the 300 miles of the San Andreas that moved in the 1906 earthquake. The part that moved is called the rupture surface. The rupture surface is approximately a rectangle with an orientation that is described by its strike, the direction it trends relative to north, and its dip, the angle it makes with the earth's surface. Which way the two sides of the fault moved is called the slip direction. The rupture's size is described by its length, the distance parallel to the earth's surface, and its width, the other dimension. "Width" is a slightly confusing word in this context, but no one has found a better one.

What happens on a fault during an earthquake is complicated because it involves two kinds of motion. At any point on the rupture surface, one side

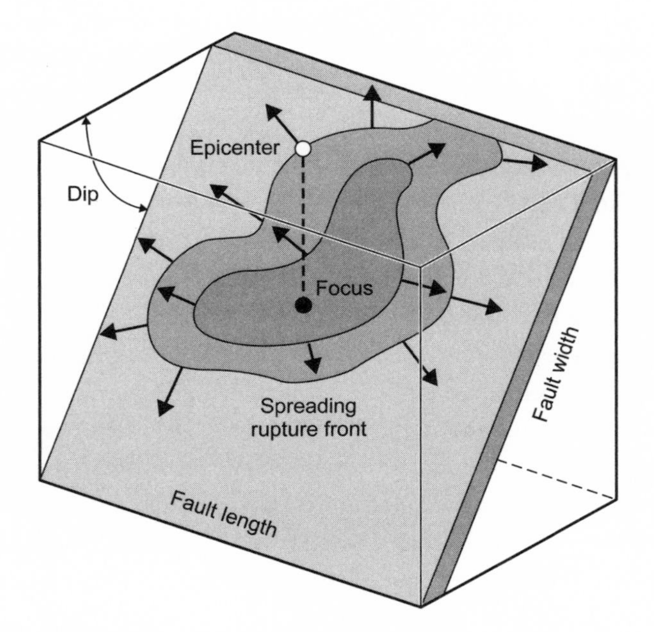

FIGURE 8.1 Geometry of the fault rupture in an earthquake.

moves relative to the other. This motion occurs first at the hypocenter and then at other points on the rupture surface as the location of rupture moves. When all the points have finished moving, the earthquake stops. A way to understand this is to think of a zipper: At each point one side moves relative to the other, and the point where this is happening moves along the zipper track.

Where Was It?

Seismologists want to know where an earthquake was to figure out which fault moved. Where the geology is known, most earthquakes happen on recognized faults, which often appear at the earth's surface. However, some faults don't show up on the surface.

In particular, the 1811–1812 New Madrid earthquakes happened on faults that are buried under thousands of feet of sediment. There was a lot of broken ground after the earthquakes, but it was caused by the strong shaking rather than motion on a fault that reached the surface. That's why ground breakage can't be used to map the fault the way geologists did after the 1906 San Francisco earthquake. That earthquake was a major earthquake, close to the earth's surface and under land. Unless all three are true, the surface doesn't rupture. Most earthquakes are small, too deep to cause a rupture visible at the earth's surface, and under the ocean.

Fortunately, the locations of earthquakes can be found using the times when the seismic waves they generate arrive at seismometers. Over the years, seismologists have used seismograms from many earthquakes to learn how fast seismic waves travel at different depths and locations in the earth. Knowing this shows how long it would take a seismic wave to get from one place on earth to a seismometer at another place. Locating an earthquake asks the question backward: Where and when must the earthquake have been for the seismic waves to get to the seismometers when they did?

Solving this question uses a method called triangulation that's like drawing circles on a map. Imagine that we already know when the earthquake happened. Because we know how fast seismic waves travel, the time when they arrived at each seismometer shows how far away the earthquake was. This means the earthquake had to have occurred somewhere on a circle around that seismometer. If we draw circles around three seismometers, the earthquake happened where the circles meet (fig. 8.2). In reality we don't know when the earthquake happened, but adding another seismometer pins down that time.

The real case is harder because the earth is a sphere and the earthquake was at some depth rather than on the surface. In the early days of seismology, earthquakes were located by drawing circles on a globe. Today, it's done by computer using the times when waves arrive at hundreds of seismometers. The process is automated, so the hypocenter is known a short time after an earthquake.

The name triangulation comes from a surveying method of finding a position using three angles, or directions. Earthquake location uses three distances, which is technically trilateration. We won't worry about the distinction because triangulation is the common name. It's like the way Americans use the term "buffalo" even though the animal that lives in the western United States is technically a bison and true buffalo live in Asia and Africa.

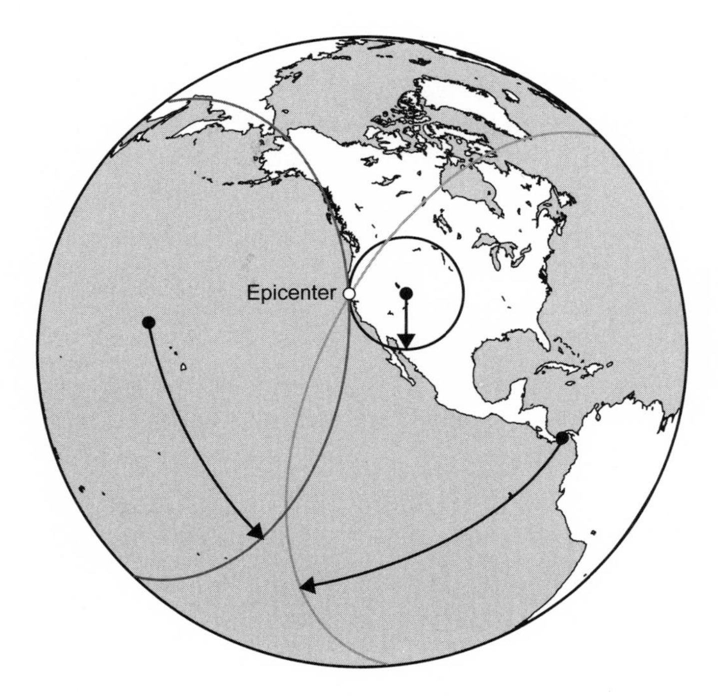

FIGURE 8.2 Locating an earthquake using the times when seismic waves arrived at different seismic stations.

The triangulation method relies on combining data from many seismometers in different places. That's why seismologists were among the first scientists to share data among different institutions around the world and are still leaders in this kind of cooperation. It's sometimes hard because geologists are an independent bunch; getting us together is like herding cats. Even so, most earthquake seismograms and other data are freely available on the Internet. This lets us work better and compare and check results. Sometimes this causes big arguments, but it gets better answers a lot faster than having people argue based on data that no one else can analyze.

As the numbers of seismometers around the world increased, the ability to locate earthquakes improved. By now, moderate earthquakes anywhere in the world are reasonably well located by global seismic networks that share data. The U.S. network is operated by the Incorporated Research Institutions for Seismology (IRIS), a consortium of universities and other institutions. IRIS was organized to raise money to build and operate seismic networks, so its name unofficially also represented Iris, the ancient Greek goddess of the rainbow, because we were looking for and found a pot of gold.

There are also regional networks of seismometers that locate smaller earthquakes in specific areas including Southern California and the New Madrid seismic zone. Thus, we have a good idea of where earthquakes have occurred in the past few decades.

It's much harder to locate earthquakes that occurred before the seismometer was invented. As we saw in Chapter 5, ideas about where the large New Madrid earthquakes happened come from combining historical accounts with the locations of smaller earthquakes that occurred in the past few decades.

How Big Was It?

Once seismologists know where and when an earthquake happened, the next question is how big it was. Earthquake sizes are given by their magnitude, which is based on the amplitude of the resulting waves as recorded on a seismogram.

Although news media focus on the magnitude itself, seismologists are more interested in what it tells about what happened during the earthquake. The magnitude depends on how large an area of the fault moved and how much it moved. When my colleague Emile Okal and I got world-wide media attention by showing that the giant 2004 Sumatra earthquake had a magnitude of 9.3 rather than the 9.0 that was first found, the number itself wasn't important. The important fact was that the area that had slipped was three times bigger than first thought. Because that part of the fault had just slipped, it would be hundreds of years before it could generate another huge tsunami like the one that had just killed more than 200,000 people.

Similarly, the reason there has been so much interest in the magnitudes of the 1811 and 1812 New Madrid earthquakes is that they give an idea about how large future earthquakes could be. As we'll see, the older estimates that the largest earthquake had a magnitude of 8.1 instead of 7.4 predicted that the ground shaking would have been five times greater and that the energy released would have been eleven times more. In turn, this predicted a higher earthquake hazard and favored a stronger level of earthquake resistant construction. Similarly, if the magnitude was 7.0 rather than 7.4, the shaking would have been three times smaller and the energy released four times less.

Although the media want to hear about magnitudes, seismologists dread talking about them because magnitudes are confusing. The problem is that there are several different magnitudes. Each magnitude depends on the formula used to calculate it from seismograms, which is called a magnitude scale. Although a "scale" sounds like the machine you use to see if a diet's working, you can't see one in a museum. A magnitude scale is just a formula.

Charles Richter introduced the earliest magnitude scale in 1935. Richter was celebrated for both his careful analysis of seismic waves and his eccentric private lifestyle, including avid nudism. His scale, developed for small Southern California earthquakes, isn't used today. Instead, seismologists use scales that can handle bigger earthquakes around the world.

Magnitude is based on the fact that the recorded amplitude of ground motion reflects the earthquake's size once it has been corrected for the fact that waves get smaller as they move away from the earthquake. Different magnitudes are labeled by "M" with a subscript showing what kind they are. A common one, the surface wave magnitude M_s, is based on the size of seismic waves that travel near the earth's surface. It's calculated using the formula

$$M_s = \log (A/T) + 1.66 \log d + 3.3$$

where A is the amplitude of the wave that is measured, T is the period of that wave—usually 20 seconds, and d is the distance from the seismometer. "Log" is the logarithm, the mathematical function that measures how many times 10 has to be multiplied to get a number. For example, 10 times 10 is 100, so the logarithm of 100 is 2. Because of the logarithm, a one unit increase in magnitude, as from 5 to 6, indicates a 10-fold increase in seismic wave amplitude.

Figure 8.3 shows how this works by comparing the motion from the 1906 earthquake recorded in Europe to the motion recorded at about the same distance from the giant 2004 Sumatra earthquake. The ground motion from the 2004 earthquake is about a tenth of an inch, corresponding to magnitude 9.3. Motion from the 1906 earthquake, which had a magnitude of 7.8, is about 10 times smaller.

The magnitude scale isn't fixed from 1–10 like Olympic diving scores. Earthquakes come in whatever size they do. Still, it works out about that way. Earthquakes range from tiny, which have negative magnitudes, to giant, with a magnitude of 9. A magnitude 9 earthquake would cause 10 million times larger ground motion than a magnitude 1. Understanding the use of the logarithm makes earthquake sizes easier to understand. There are also words to describe magnitudes. Magnitude 3 earthquakes are minor, 4 are light, 5 are moderate, 6 are strong, 7 are major, and 8 or higher are great. The biggest since the invention of the seismometer was the magnitude 9.5 earthquake in 1960 along the Chilean coast.

Magnitudes like the surface wave magnitude have a major weakness. Because they were defined before anyone knew much about earthquakes, they

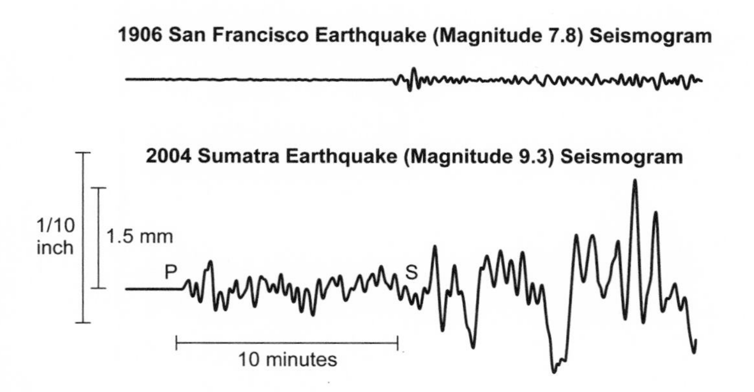

FIGURE 8.3 Comparison of seismograms for the 1906 San Francisco and 2004 Sumatra earthquakes. (Richard Aster)

don't directly reflect what actually happened during the earthquake. Seismologists want to know how large an area of the fault slipped and how far it slipped. This is important for many reasons, including getting an idea of when the fault will slip again.

To get around these problems, seismologists use the "seismic moment," calculated by measuring the energy in the longest periods of the seismogram. The seismic moment is defined as

$$M_o = \text{fault rigidity} \times \text{fault area} \times \text{slip distance.}$$

Because the seismic moment depends directly on what happened on the fault during the earthquake, measuring the moment lets us study what happened and use other observations to check the result. As always in science, doing things in different ways and comparing the answers is the best way of checking how accurate the answers are. We have a good idea of the fault's rigidity, or strength, from both earthquake studies and lab experiments on rocks. The area of the fault can be estimated from seismograms. Another way is to look at the locations of aftershocks, which are smaller earthquakes that happen after a big earthquake. Because these are on and near the fault plane, the aftershocks map out the area of the fault that moved. Dividing the measured seismic moment by the rigidity and fault area shows how far the fault slipped. If the earthquake ruptured the earth's surface, these values can be checked by measuring the length of surface rupture and the amount of offset across it.

Although seismologists like seismic moments, they're hard to explain to the media and public. They come in units like 10^{29} dyne-cm that just don't have the same ring as magnitude 8.6. To get around this, Tom Hanks (no relation to the actor) and Hiroo Kanamori came up with a clever solution. They developed the moment magnitude M_w that is calculated from the seismic moment using the relation

$$M_w = (\log M_o / 1.5) - 10.73.$$

The constants in this equation were picked so that the moment magnitude approximately matches other magnitudes, so the moment magnitude is now called "the" magnitude.

The terminology is important for New Madrid because the first studies of the 1811–1812 earthquakes took place before moment magnitudes were invented. Different studies of the historical data gave their results using different magnitude scales. Today, all the studies use moment magnitude.

Why Magnitude Matters

The moment magnitude lets us relate the amount of ground motion to what happened on the fault. To see how this works, look at the comparisons in figure 8.4. The boxes show the area of the fault that broke in each earthquake, which is the length along the fault times the fault width.

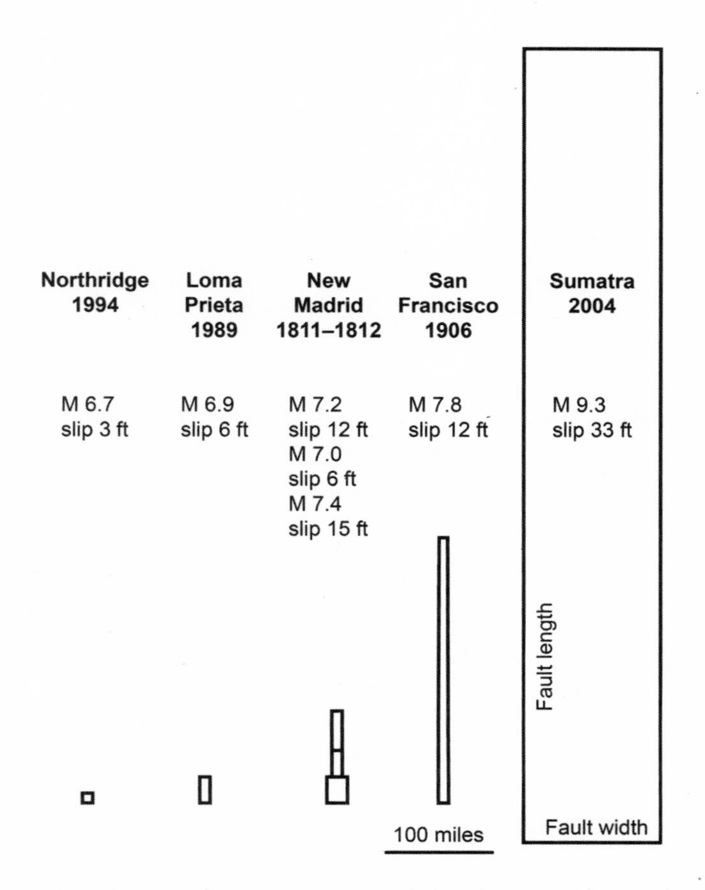

FIGURE 8.4 Comparison of rupture areas and slip distances for earthquakes with different magnitudes.

The smallest earthquake shown is the magnitude 6.7 1994 Northridge, California, earthquake. In that earthquake, an area about 10 miles on a side slipped an average distance of about 3 feet. Although the earthquake wasn't that big, it occurred below the heavily populated Los Angeles area and so caused 58 deaths and an amazing $20 billion worth of damage. The next earthquake shown is the magnitude 6.9 Loma Prieta earthquake, in which an area of the San Andreas fault 25 miles long and about 10 miles wide slipped about 6 feet. These two are a lot smaller than the 1906 earthquake. As we saw in Chapter 6, an area of the San Andreas about 300 miles long and about 10 miles wide slipped about 12 feet.

Although in California people often talk about the next big San Francisco earthquake—the recurrence of the 1906 one—as "the big one," the 2004 Sumatra earthquake was almost unbelievably bigger. In it, an area about 750 miles long and 125 miles wide—about the same size as the whole state of California—slipped about 33 feet. The Sumatra earthquake is the third largest since the invention of the seismometer, after the 1960 Chilean and 1964 Alaskan earthquakes.

Even using seismograms and other modern data, the values found for magnitude, fault area, and slip distance have uncertainties. The final value that's reported is an average of results from seismometers in different places. These results differ because of differences in how well seismic waves travel through the earth. In addition, different methods give somewhat different results. Thus, studies of the Loma Prieta earthquake find magnitudes that vary by 0.2 units. We could say the magnitude was 6.8, 6.9, or 7.0. Older data cause even greater uncertainties. The magnitude of the 1906 San Francisco earthquake was once thought to be 8.3 but is now thought to be about 7.8, and the fault length has been estimated as anywhere between 186 and 310 miles. Fault widths are especially hard to get good values for, in part because from the earth's surface geologists don't see the bottom of a fault.

Comparing different earthquakes shows that bigger earthquakes happen when the ground slips for a longer distance on a larger fault plane. This fact is important for understanding the strong ground shaking that happens close enough to the earthquake to cause damage. Earthquake magnitude is based on the fact that the shaking is greater for larger earthquakes. In addition, the ground shakes for a longer time in larger earthquakes, as shown in figure 8.5. That's because the faulting starts at the

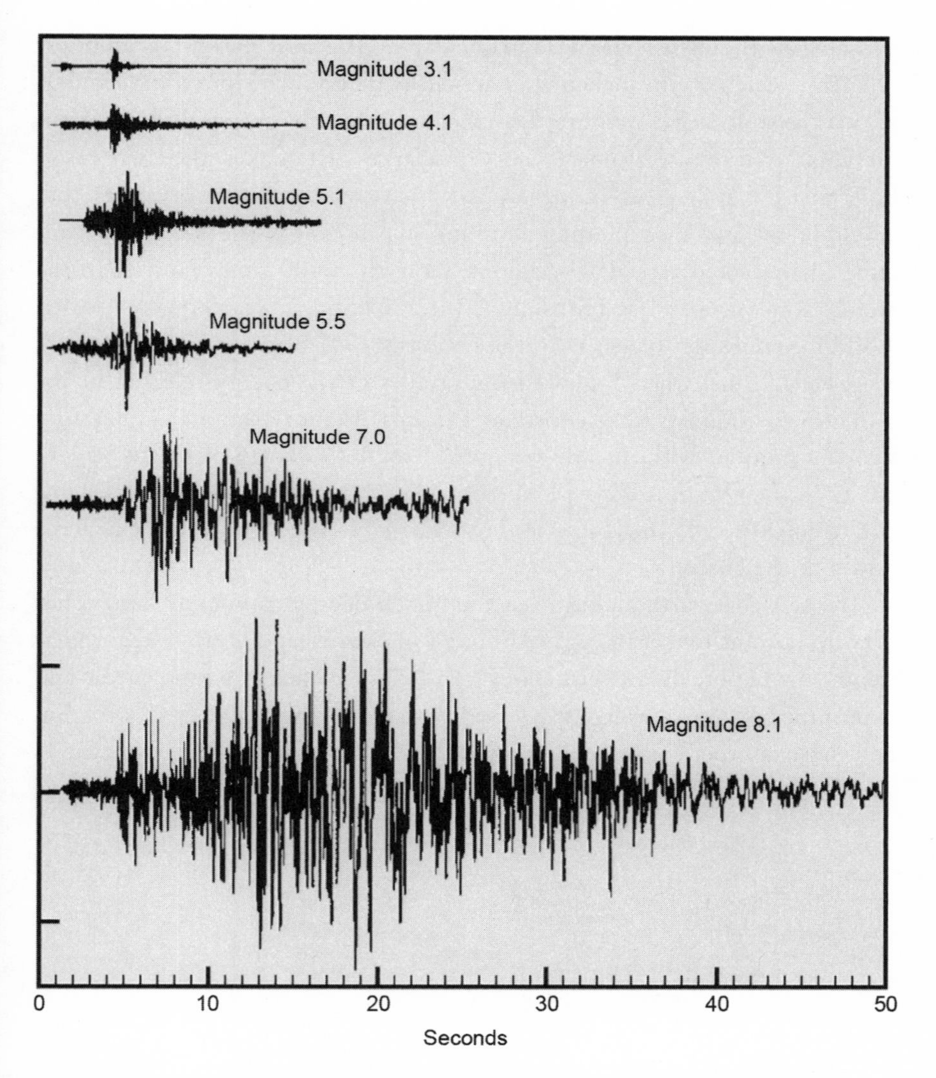

FIGURE 8.5 Comparison of the duration of shaking for earthquakes of different magnitudes. (University of Nevada, Reno)

focus but takes a while to move along the fault. Seismic waves radiate from the entire fault, so a longer fault produces a longer duration of shaking. This is important for designing buildings because a building that survives a short duration of shaking might not survive a longer one unless it's designed to.

Large earthquakes can cause major damage because enormous amounts of energy stored on the locked fault are released and travel into the surroundings as seismic waves. Figure 8.6 compares the energy released in earthquakes to that released in other ways. For large earthquakes, the energy is so huge that it can be given in megatons, which are the amount of energy that would be released by exploding a million tons of TNT. The 1906 San Francisco earthquake released 7 megatons, more than 400 times more than the atomic bomb dropped on Hiroshima. The 2004 Sumatra earthquake released 1,300 megatons, far bigger than the largest nuclear bomb ever exploded (58 megatons). Thus, when I showed our studies of the energy released in the Sumatra earthquake to scientists at the Los Alamos National Laboratory who are familiar with nuclear weapons, I reminded them of Darth Vader's advice in the film *Star Wars,* "Don't be too proud of this technological terror you've constructed. The ability to destroy a planet is insignificant next to the power of the Force."

The fact that earthquakes release as much energy as nuclear bombs has been important for seismology. Although nuclear weapons are tested underground in remote areas, seismology can detect a nuclear weapons test and determine how much energy it released. As a result, the U.S. defense program and those of other countries supported the growth of seismological networks

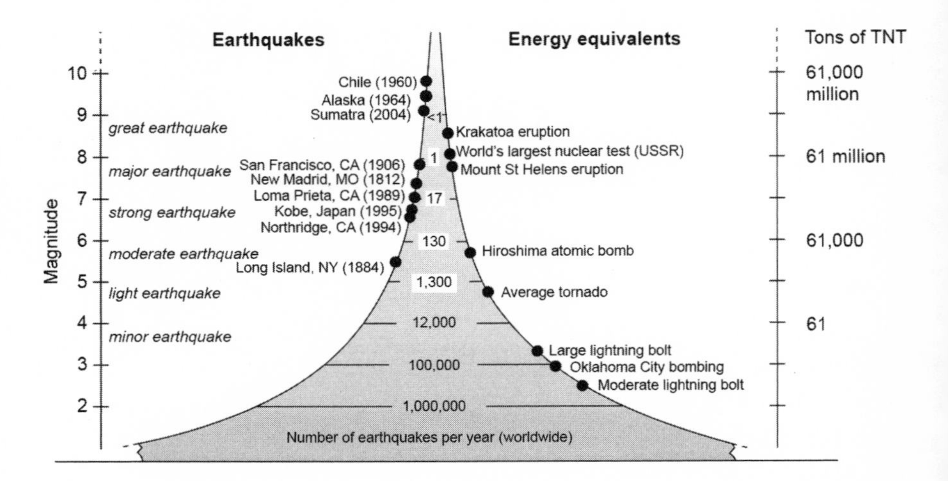

FIGURE 8.6 Comparison of earthquakes with different magnitudes in terms of how often they happen and the energy they release. (Stein and Wysession, 2003)

and conduct active monitoring programs. These seismologists do research without the "publish or perish" rules of university science because often their most interesting results can't be published. One of their biggest jobs is distinguishing between nuclear tests and earthquakes. It helps that bureaucratic instincts schedule most tests exactly on the hour, but even without this clue, seismological methods can tell the difference. These methods have allowed treaties that limit the size of nuclear tests.

An important point is that big earthquakes release much more energy than small ones. A one-unit increase in magnitude, for example from 5 to 6, increases the wave energy by a factor of about 32. Hence, a magnitude 7 earthquake releases 1,000 times more energy than a magnitude 5. This is why the hazard from earthquakes is mostly from large earthquakes, typically with magnitudes greater than 6. As a result, when a magnitude 5 earthquake like the one discussed in the past chapter occurs every 20 years or so in the central United States, it's a great news story, but seismologists don't get too excited.

Big Earthquakes Are Rare

Big earthquakes are less common than small ones. Earthquakes of a given magnitude occur about 10 times more frequently than ones a magnitude larger, as shown in figure 8.6. For the whole world in a typical year, there is about one magnitude 8 earthquake, about 17 magnitude 7s, about 130 magnitude 6s, 1,300 magnitude 5s, and so on. The exact numbers vary from year to year, but the factor of 10 relation is clear. Most of the energy is released in the few largest earthquakes. Often, the biggest earthquake in a given year releases more energy than all of the others that year. That's why although there are millions of earthquakes every year, only a few large ones do any harm.

These numbers give a reality check when people claim to be predicting earthquakes. Usually, they do no better than just guessing based on the average. For example, there's a magnitude 7 earthquake somewhere in the world about every month and a magnitude 6 about every three days. In Chapter 2, we mentioned that as part of his Midwest earthquake prediction, Iben Browning claimed to have predicted the 1989 Loma Prieta earthquake. He'd actually said that near that date there would be an earthquake somewhere in the world with a magnitude of 6. Because that was a safe but useless bet, seismologists weren't impressed.

For the same reason, statements about the 1811–1812 New Madrid earthquakes that refer to them as "the most powerful series of earthquakes ever known on earth," like in the video *Hidden Fury,* are pretty silly. There are about 15–20 such earthquakes every year.

The factor of 10 relation between the number of earthquakes and their magnitude works in individual earthquake zones as well as around the world. For example, in the past 1,300 years Japan has had about 190 magnitude 7 or greater earthquakes and 20 magnitude 8 or greater. Similarly, since 1816 Southern California has had about 180 magnitude 6 or greater earthquakes and 25 magnitude 7 or greater.

Using the number of small earthquakes is one way to estimate about how often a large earthquake will happen in places like New Madrid, where large earthquakes are rare so our record of them isn't good. Since 1975, seismometers have recorded the number of earthquakes there. Before then, there were historic earthquakes, whose magnitudes have been estimated. Dividing the number of earthquakes above a given magnitude by the time period gives an average number per year. The reciprocal of that number is the average time between earthquakes of that size.

Figure 8.7 shows the results. There's a magnitude 5 or greater earthquake somewhere in the New Madrid seismic zone about every 20 years. Similarly, there's a magnitude 6 or greater earthquake about every 175 years. The exact number depends on the precise magnitude values used and how large an area is considered to be the New Madrid seismic zone. These numbers are for the main part of the zone, spanning 35°–38° north latitude and 88°–91° west longitude. A graph shows that these numbers are close to a straight line because of the factor of 10 relation.

These graphs are called frequency-magnitude or Gutenberg-Richter plots. Researchers plot the rate of small earthquakes and extend the line to estimate about how often a large earthquake will happen. In this case, the dashed line predicts that a magnitude 7 or larger earthquake should happen about every 1,500 years. Other methods give different values, as we'll see.

Patterns like big earthquakes being much less common than small ones happen in many other situations. There are many small storms but few hurricanes, many small towns but few huge cities, lots of middle income people but few billionaires, and so on. Situations like these create what is called the

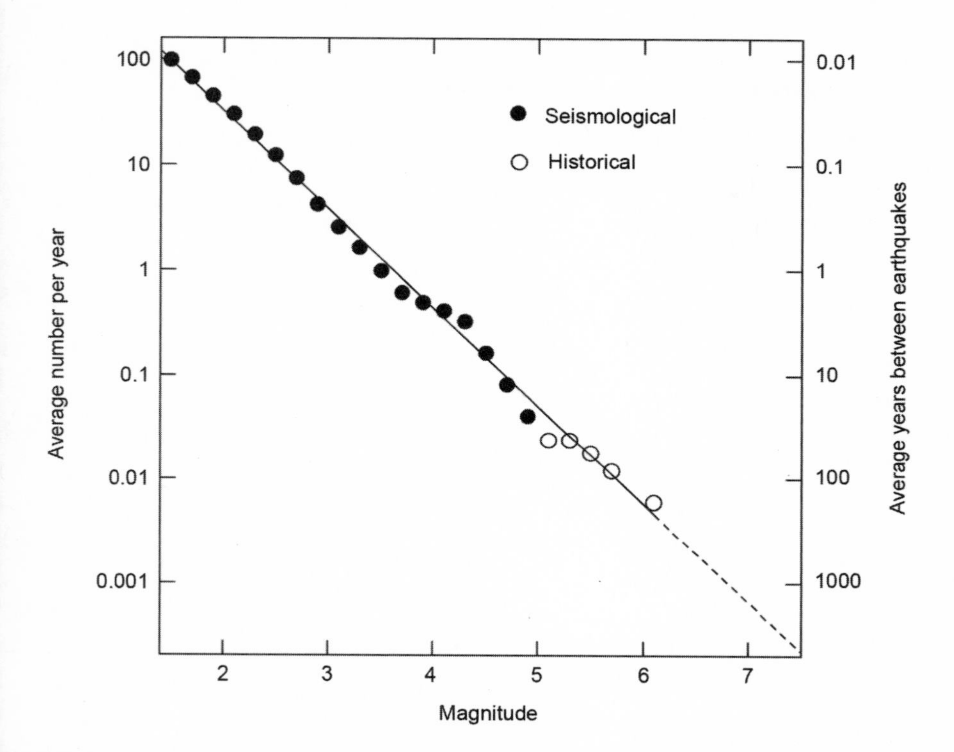

FIGURE 8.7 Average number of earthquakes per year in the New Madrid seismic zone with magnitude larger than or equal to a given value. (Stein and Newman, 2004)

80-20 rule: 80% of some effect comes from 20% of the causes, 80% of wealth is owned by 20% of the people, 80% of a business' sales come from 20% of customers, 80% of horse races are won by 20% of jockeys, and so on. It's even been suggested that people wear 20% of their clothes 80% of the time.

How Did the Fault Move?

Seismograms can show how a fault moved in an earthquake. The first step is to find the fault geometry, given by the fault strike and dip directions and the direction of motion between the two sides. Figure 8.8 shows some possibilities. All four of these faults strike to the north, but the top two dip vertically into the earth whereas the bottom two dip 45° to the east.

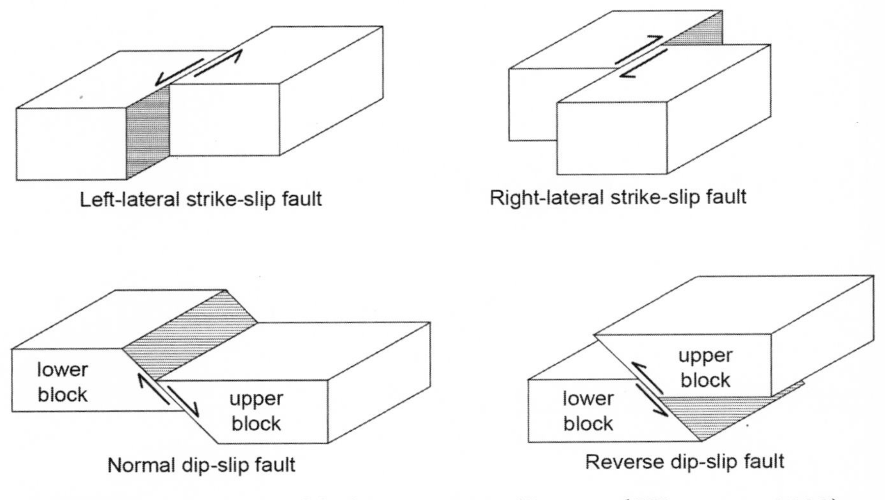

FIGURE 8.8 Comparison of fault geometries. (Stein and Wysession, 2003)

The top two faults show strike-slip motion, in which the two sides of the fault slide horizontally by each other. The one on the left is called left-lateral strike-slip motion because if you look across the fault you see that the other side moved left. If the fault moved the other way, it's called right-lateral strike-slip.

The bottom two faults show dip-slip motion, in which the two sides move vertically. When the upper block slides downward, it's called normal faulting. The opposite case, in which the upper block slides upward, is called reverse faulting. Many earthquakes show combinations of strike-slip and dip-slip motion.

The kind of faulting that happens shows what is happening to rocks near the fault. If they are being pulled apart, normal faulting happens. If they are being pushed together, reverse faulting happens.

When an earthquake ruptures the earth's surface, the rupture shows a lot about the type of faulting. When there's no surface rupture, as in the New Madrid seismic zone, the fault geometry and direction of motion are found from the three-dimensional pattern of the seismic waves that the motion produced. Although the math is complicated, the idea is simple.

Figure 8.9 shows how it works for a vertically dipping right-lateral strike-slip fault. U.S. seismologists often draw a fault this way because that's what the San Andreas is. As the material on one side moves horizontally with re-

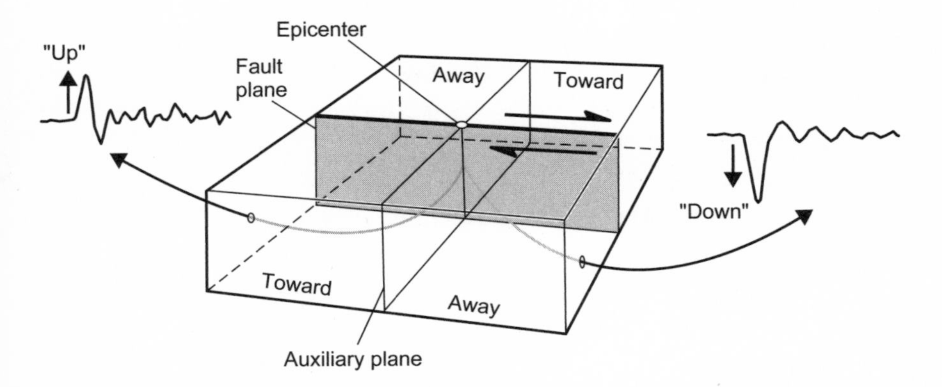

FIGURE 8.9 Finding fault geometry from the polarity of P waves. (Stein and Wysession, 2003)

spect to that on the other side, the motion generates seismic P waves that propagate away in all directions. These waves have different shapes. The most useful feature is the direction of the first ground motion, called the polarity. In some directions, the ground first moves toward a seismic station, whereas in other directions the ground first moves away from a station.

This motion makes the seismograms look different at different stations, which can be divided into four zones or quadrants. In two quadrants, the first ground motion is toward the station, whereas in the other two quadrants, the first ground motion is away from the station. Because the seismic waves go down into the earth as they leave the earthquake, turn at depth, and come up again from below at a distant seismographic station, the first motion is upward in a "toward" quadrant and downward in an "away" quadrant.

The quadrants are picked using seismograms from stations at different directions from the earthquake. Next, the nodal planes—the fault plane and an auxiliary plane perpendicular to it—are identified because in these directions the first motion changes polarity from "up" to "down." Using additional data can often tell which plane was the actual fault.

This method combines data from seismometers in different places. It's easy to install seismometers the wrong way, like switching east-west and north-south components or switching wires so "up" motion is recorded as "down." As a result, much effort goes into installing and testing seismometers to avoid wrong "discoveries."

There's more information in the waves than just the polarities. The pulses' sizes also give information about the fault geometry because they're small in the directions of the nodal planes, across which they change polarity, and largest halfway between the planes. Put another way, the fault radiates P waves best at angles of 45° to the fault plane. You hear a similar effect when you turn a radio around to get better reception. What's happening is that the antenna in the radio is more sensitive in some directions than others.

The pulses' shapes are used to make detailed maps of what happened on the fault during an earthquake. They show the amount of slip that occurred, the size of the area that slipped, and the history of the slip. For example, the pulses are taller and narrower on seismograms in the direction the rupture moved because the wave energy "piles up." The principle that the wave's period is shorter in the direction that the source moves is called the Doppler effect. Police radar guns use it to measure the length of the pulse that bounced off a car to see how fast it's going.

Thinking about fault areas brings up the question of why earthquakes have the sizes they do. Why did the 1906 earthquake break only 300 miles of the San Andreas? Why does faulting only go so deep into the earth? Understanding how large an earthquake can occur on a particular fault is important for understanding the earthquake hazard there.

Seismologists don't understand the first question, though there are some ideas. The biggest possible earthquake on a fault is one in which it all breaks at the same time, but it might also break in smaller pieces at different times. The earthquake history gives a guide because often earthquakes rupture the same piece of the fault, called a fault segment, over and over again. Hence, when a 1906-style earthquake happens again, it's expected to rupture about the same portion of the San Andreas as last time. Similarly, the segment to the south that broke in a similar-size earthquake in 1857 is expected to rupture again. However, because bigger earthquakes are rarer than smaller ones, the fact that we've seen only smaller ones doesn't mean bigger ones can't happen. The giant 2004 Sumatra earthquake ruptured a much longer length of fault than was expected from the known earthquake history, and so generated a much bigger tsunami than expected. Geological studies since then show the last such earthquake was about 600 years ago. This example shows that ideas based on the short earthquake history can be misleading.

The temperature of the rocks controls the maximum depth of faulting. Lab experiments show that when rocks get hotter than about 600° Fahrenheit, their friction gets much less. Instead of sticking for a while and then suddenly sliding, like the soap bar on the yoga mat, they just slide. It's hotter deeper in the earth, so below about 15 miles, the rocks of the continents aren't involved in elastic rebound, and earthquakes don't go deeper. Because faults like the San Andreas dip vertically into the earth, an earthquake rupture on them can't get very deep. In contrast, the 2004 Sumatra earthquake was much bigger in part because it happened on a fault that dips at a shallow angle from the horizontal, about 20°, so it goes a long way down before the rock gets too hot to rupture.

What Happened in the New Madrid Earthquakes

Figuring out what happened in the 1811–1812 New Madrid earthquakes is more complicated than for modern earthquakes because seismometers hadn't been invented. Instead, researchers combine indirect methods with what's known about modern earthquakes.

Because the magnitudes can't be measured, they're estimated by a complicated process, each step of which has a lot of uncertainty. As we saw in Chapter 5, reports of shaking and damage are used to assign numerical values to the intensity of shaking. The pattern of the intensities—how large an area different values cover—is then compared to that for modern earthquakes with known magnitudes. Because seismic waves travel better in some places than others, this comparison should be done for other large earthquakes in eastern North America, but these are rare. The best comparison is to the 1929 magnitude 7.2 earthquake on the Grand Banks of Newfoundland, which seems to have been about the same size.

A complexity is that the faults that caused the big 1811 and 1812 New Madrid earthquakes are mostly buried. Thus, researchers assume that these faults are indicated by the locations of recent smaller earthquakes (fig. 8.10), many of which are aftershocks that happened on or near these faults. The small earthquakes map distinct faults, with the Cottonwood Grove fault offset from the North New Madrid fault by a "jag" at the Reelfoot fault.

Without seismograms, researchers can't use modern methods to tell how the faults moved in the big earthquakes. However, some ideas come from combining the historic accounts with studies of seismograms from smaller modern earthquakes. Figure 8.10 shows one of the scenarios that has been

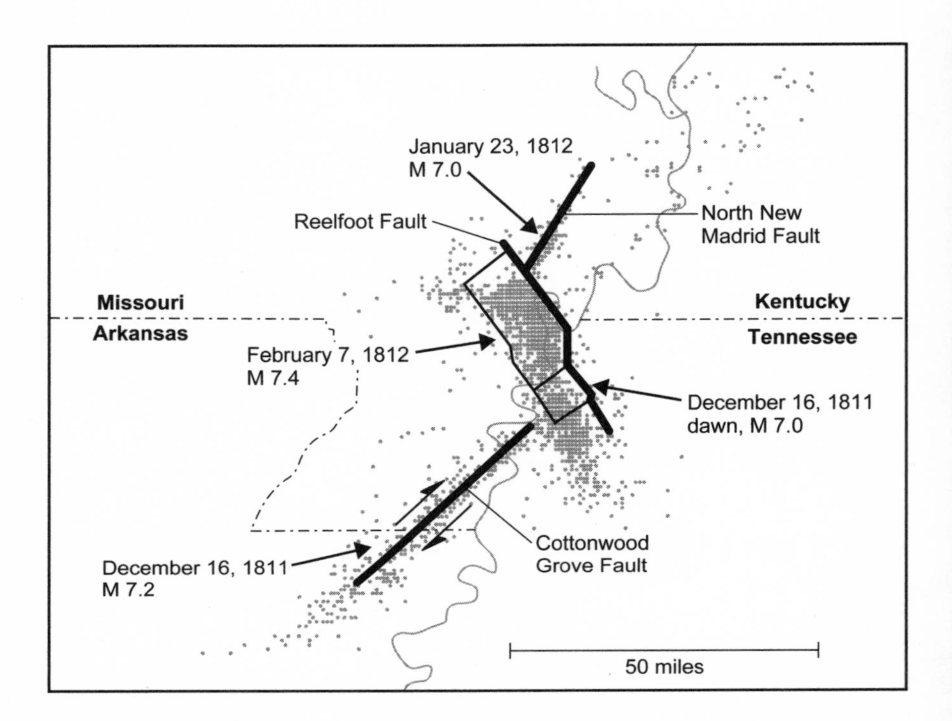

FIGURE 8.10 A scenario for the locations, fault lengths, and magnitudes of the largest New Madrid earthquakes of 1811 and 1812. Dots show the epicenters of recent small earthquakes. (Hough, 2004)

proposed. In it, the magnitude 7.2 December earthquake involved about 12 feet of motion on a part of the Cottonwood Grove fault 40 miles long and 10 miles wide. The large morning aftershock occurred on the east end of the Reelfoot fault, but its fault dimensions aren't constrained. The January earthquake was a magnitude 7.0 event with about 6 feet of slip on a piece of the North New Madrid fault 25 miles long and 10 miles wide. The February shock was the biggest, magnitude 7.4, with about 15 feet of slip on a part of the Reelfoot fault 25 miles long and 20 miles wide. Of the three largest earthquakes, the scenario for the January one is considered the most uncertain. Adding the three big earthquakes, as shown in figure 8.4, and the December aftershock together gives a total magnitude of 7.6, almost as big as the 1906 San Francisco earthquake. The smaller magnitudes coming out of the recent study would make New Madrid significantly smaller.

In this scenario, there was an important difference between the three big shocks. The December and January earthquakes involved right-lateral strike-slip motion on vertically dipping faults. That would have compressed the rock on the Reelfoot fault, where the February earthquake occurred. The February earthquake involved reverse faulting, which raised the ground on the south side of the fault and so disrupted the flow of the Mississippi River. This fault dips at a shallow angle to the south, which is why the small earthquakes on it show up as a broad zone in figure 8.10. Because of the shallow dip, the February earthquake had a wider fault area than the other two earthquakes. Thus, even though the February earthquake might have had the shortest fault length, it was the biggest earthquake.

This scenario seems plausible given what's known. However, given the uncertainties involved, seismologists will continue debating exactly what happened for years to come—and there will probably never be a final agreement.

Chapter 9

Plate Tectonics Explains (Most) Earthquakes

> Once the idea of plate tectonics had been clearly formulated, it was amazing how quickly and how neatly a large body of seismological data fell into place.
>
> —Paleomagnetist Allan Cox, 1926–1987

In the past two chapters, we've seen what seismic waves are and how earthquakes produce them. These results let us understand what happened in the 1811–1812 New Madrid earthquakes. The next question is what causes earthquakes. Why are faults where they are? What makes them move? Answering these questions is important for trying to decide if another big New Madrid earthquake is likely any time soon.

Developing the Idea

After coming up with the brilliant idea that elastic rebound—the release of motion stored on the locked fault—caused the 1906 San Francisco earthquake, geologists studying it hit a dead end. They thought that the motion between the two sides of the fault happened within 30 miles of the fault and had no idea what caused it.

With 20/20 hindsight, the answer was staring them in the face, but they just didn't see it. That's often the case in science, whether in studying a huge problem like what causes earthquakes or a smaller problem like the New Madrid earthquakes. We go so far, get stuck, and can't see what to do. Then someone else points out a simple answer that we somehow missed. It's fun when we're the people pointing out the answer and embarrassing when someone does it to us.

The answer to what causes earthquakes stares out from every world map. Geologists love maps. We cover our office walls with them and get some of our best ideas from them. The geologists of 1906 knew how nicely the coasts of South America and Africa fit together. This had been noted since the 1600s, but nobody realized that it was the key to understanding how the earth works.

The breakthrough came from Alfred Wegener, a German meteorologist who is to geology who Newton is to physics and Darwin is to biology. In 1915, while recovering from being wounded in World War I, he published a book called *The Origin of Continents and Oceans* that changed geology. He offered a variety of evidence beyond the fit of the coasts of South America and Africa to support the theory he called continental drift. One line of evidence used the fact that glaciers moving on the earth's surface leave characteristic rock deposits at the glaciers' edges and scratches on rocks showing the direction the glaciers moved. Glaciated areas of the southern continents can be fit back together, implying that these continents had been together near the South Pole. Another line of evidence used fossils of species that are now found in areas far apart to argue that these areas must have been joined in the past. Wegener argued that 300 million years ago, the continents were joined in a supercontinent called "Pangaea," which means "all land" in Greek, and later drifted apart (fig. 9.1). This idea explained the glacial data, the fit of the continents, and the fossils.

What happened next is one of every geologist's favorite stories. Some geologists who were familiar with rocks in the southern hemisphere accepted Wegener's idea. However, most in the northern hemisphere were highly critical. There were many objections. At a gut level, many geologists thought the whole argument was weak, sloppy, and didn't make sense. Rollin Chamberlin of the University of Chicago said, "Wegener's hypothesis in general is of the footloose type, in that it takes considerable liberty with our globe, and is less bound by restrictions or tied down by awkward, ugly facts than most of its rival theories." Even worse, "If we are to believe in Wegener's hypothesis we must forget everything which has been learned in the last 70 years and start all over again." Of course, that's what science is all about!

The argument over continental drift was an example of one of science's toughest challenges. Usually, science makes slow progress by doing routine things. To make major progress, scientists need to identify the weaknesses in current ideas, question these ideas, and develop new ones. We need to question

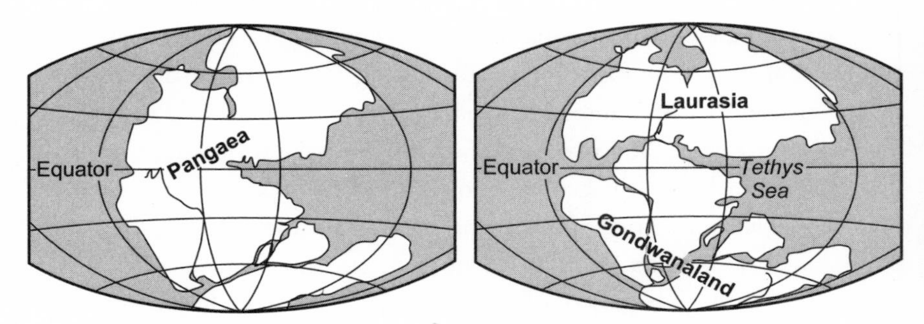

225 million years ago

200 million years ago

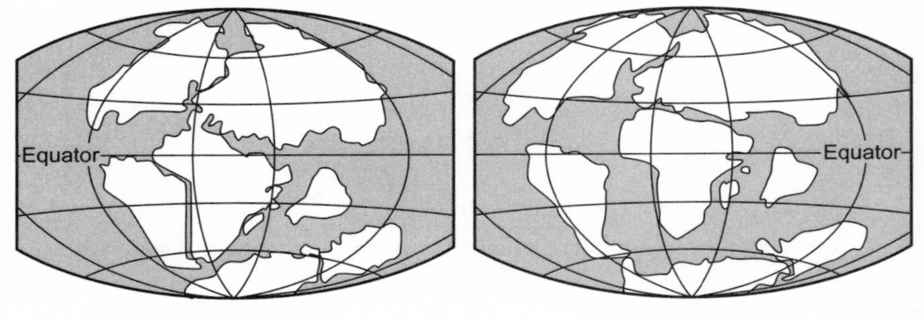

135 million years ago

65 million years ago

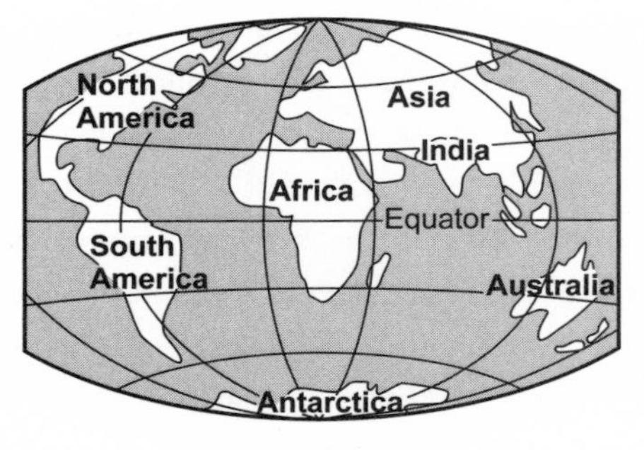

Present Day

FIGURE 9.1 Reconstruction of Pangaea and the motions of the continents since its breakup. (After USGS)

the conventional wisdom that we've been told is true and push past it. This means discarding old ideas, even if we've always believed them.

Professors try to teach students to "think outside the box," often using Wegener as a role model. We all have favorite quotes saying the same message in different words. One comes from Nobel Prize winner Peter Medawar. "I cannot give any scientist of any age better advice than this: the intensity of the conviction that a hypothesis is true has no bearing on whether it is true or not. The importance of the strength of our conviction is only to provide a proportionally strong incentive to find out if the hypothesis will stand up to critical examination."

Still, scientists are human. We cling to what we're familiar with and think we know, because that's easy. Groupthink encourages us to stay with the pack. Thus, given new ideas or new data, how should we deal with them? We don't want to jump on a new idea if it's wrong or reject it if it's right. Often, inertia wins, and new ideas take a long time to get accepted. That's not all bad because the struggle for acceptance filters out most bad new ideas and sharpens good new ones. During debates, it's hard to tell which new ideas are good. Eventually, the good ideas win, if only because generations change. My colleague Donna Jurdy jokes that science advances one funeral at a time.

The biggest weakness in Wegener's idea was that it didn't explain how continental drift worked. Scientists thought the earth was too solid for continents to move. How could they move across the sea floor? What possible force could move them? Geologists debated these problems, and most decided that continental drift couldn't work. If it couldn't work, it hadn't happened. In hindsight, this was a mistake. Scientists often understand what happened before they know why. For example, the elastic rebound theory of earthquakes was accepted based on observations even though no one knew why the sides of the fault moved.

Most geologists ignored the fit of the coastlines and rocks. Paleontologists found an alternative to explain how fossils matched across the ocean. The idea was that land bridges had connected continents, the way Panama connects North and South America. These bridges had come up and then sunk. The problem was that there wasn't any reason to believe in these bridges, except for the match of the fossils, and no one had any idea how they could rise and sink. In hindsight, the question was really which idea was more likely, continental drift or land bridges?

Scientists often have to decide between alternative explanations. One often-suggested approach is to use a principle called Ockham's razor, introduced by the fourteenth-century English philosopher William of Ockham. The idea is that the simplest solution—the one that needs the fewest assumptions—is best. The problem with using this principle is that both sides of an argument think their explanation is simpler. In this case, was continental drift or land bridges simpler? The two sides couldn't agree, so the debate went on and didn't get anywhere for 30 years. Continental drift faded from most geologists' thinking.

As often happens when science is stuck, a new kind of data changed everything. The earth has a magnetic field, like that of a giant magnet. The field is shown using lines of force that indicate the direction a compass needle points (fig. 9.2). A compass needle is a magnet, so it lines up with the earth's field, pointing north. Lines of force come out of the south magnetic pole and into the north magnetic pole, so they're horizontal at the equator and dip up or down anywhere else. This means that a vertical compass shows the latitude—how far north or south it is.

In the 1950s, studies of the earth's magnetic field made amazing discoveries. These used the fact that some rocks record the earth's magnetic field. This happens because many volcanic rocks contain minerals whose grains form tiny magnets. When hot lava pours onto the earth's surface, the earth's magnetic field lines up the minerals, like a compass needle. When the lava cools into solid rock, the magnetic minerals are "frozen" and can't move. Curiously, some volcanic rocks had magnetic minerals that didn't line up with the earth's field. Either the rocks had moved or the earth's field had changed since the time the rock cooled. It turned out both had happened.

To make sense of this, geologists turned to a new technology, radioactive age dating. Radioactivity is the process by which an atom of one element decays, or changes, into another. Once the rate of decay has been measured, comparing the amounts of the original "parent" and the resulting "daughter" elements shows how long the decay has been going on. For cooling lava, the "clock" starts when the rock becomes solid, so none of the daughter element produced by the decay can escape. As a result, radioactive dating measures the age when the rock became solid.

Dating rocks and measuring their magnetization showed that at times in the past the earth's magnetic field was opposite to today's. The north magnetic pole, which is now near the North Pole, was then at the South Pole, so

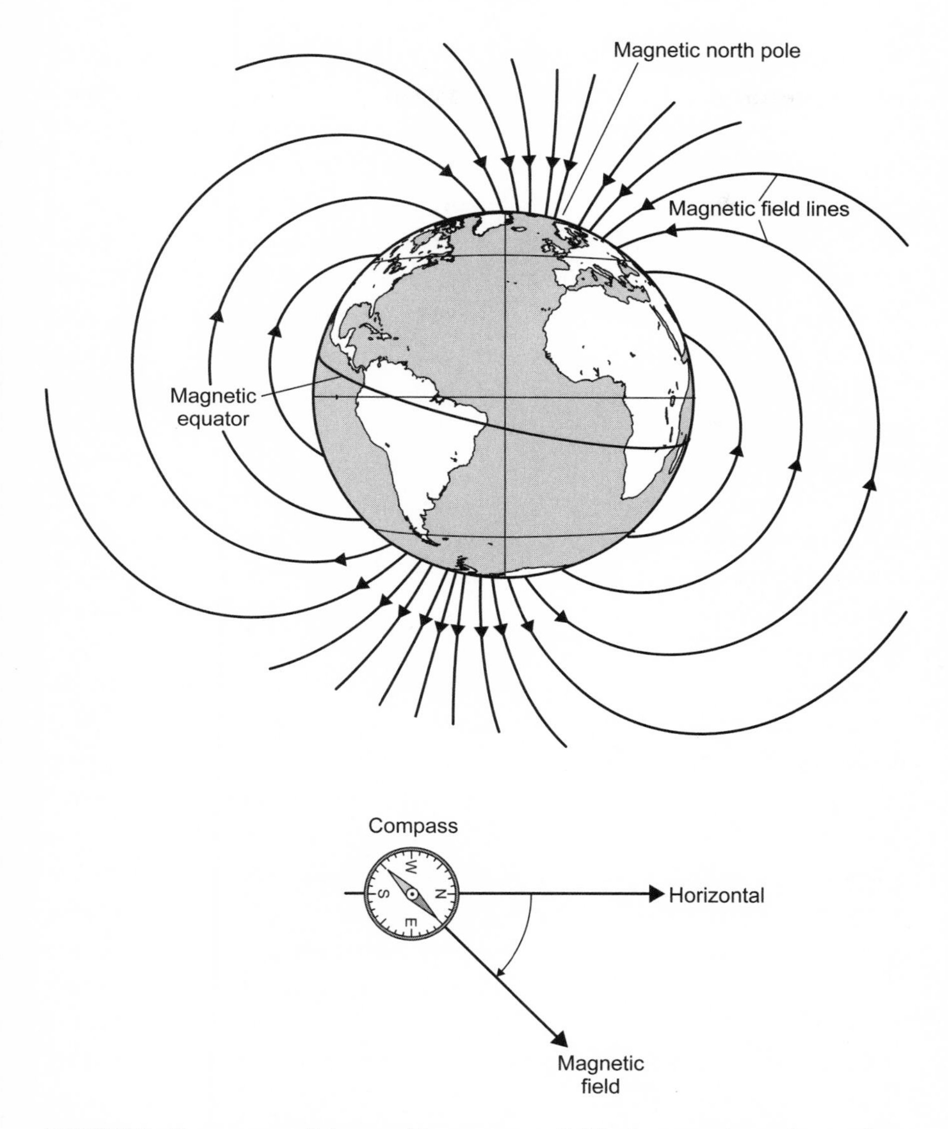

FIGURE 9.2 Geometry of the earth's magnetic field, showing how the angle it dips from the horizontal depends on latitude.

a compass would point south instead of north. A history of the earth's magnetic field, called the paleomagnetic time scale, was worked out. The time scale showed that the field reversed at irregular intervals about 200,000 years apart. This last happened about 780,000 years ago. These studies are called paleomagnetism because "paleo" is the Greek word for "old." Geology is full of "paleo" words, like paleontology, the study of ancient life. We'll be talking soon about paleoearthquakes.

Paleomagnetic studies also showed that rocks of the same age must have moved relative to each other. In particular, rocks from opposite sides of the Atlantic must have been next to each other millions of years ago. Paleomagnetism, soon jokingly called "paleomagic," proved that continental drift had happened. It gave much of the data that went into the maps of where the continents had been. It soon also showed that all parts of the earth's outer shell, not just the continents, were moving.

Within a few years, from about 1963 to 1970, continental drift went from a strange idea to accepted. It became the core of plate tectonics, the most important concept in geology. I was in the first generation of geologists to learn plate tectonics in beginning college classes. Thus, although young scientists have healthy skepticism about older ones, it was amazing to meet people who had discovered so much so recently. For example, when I was a post-doctoral fellow at Stanford University, the earth sciences dean was Allan Cox, one of the discoverers of the paleomagnetic time scale.

Plate Tectonics

Plate tectonics (from the Greek word "tecton," meaning "builder") describes the earth's outer shell as made up of about 15 major plates and about 40 minor plates. Plates move relative to each other at speeds of a few inches per year, about as fast as fingernails grow. Continents don't move through the rocks of the ocean floor. Instead, they move with the ocean floor because they're part of plates.

Plates are made up of cold, strong rocks that form the earth's outer shell, which is about 60 miles thick. This shell is called the lithosphere, from the Greek word "lithos," meaning "rock." The plates drift over warmer and weaker rocks below called the asthenosphere. The lithosphere and asthenosphere are units defined by how strong they are and how they move. What's a bit confusing is that the lithosphere includes both the crust and part of the upper mantle, which are regions identified by differences in seismic velocity, as we saw in Chapter 7.

Plates are pretty rigid, which means that their insides don't deform much. Instead, almost all the geological action happens at the boundaries between plates, like the San Andreas fault. These are where most earthquakes and volcanoes are, where mountains get built, and so on. Plates are like ice chunks floating around on water, sliding by and banging into one another.

What happens at the boundary between two plates depends on the motion between them. At divergent boundaries, plates move away from each other, whereas at convergent boundaries, they move toward each other. At the third type, transform fault boundaries, plates slide by each other.

The different kinds of boundaries are simplest when they are in the rocks under the oceans (fig. 9.3). Divergent oceanic boundaries, known as mid-ocean ridges or spreading centers, are long lines of underwater volcanoes. Here, hot rock upwells, cools, and is added to the two plates. The material continues cooling as the plates move away from the ridges. Over time, the "new" parts of the plates eventually reach convergent boundaries known as subduction zones or trenches. Here, plates are consumed as they descend back into the mantle. As they heat up, fluids are released that rise and make the mantle above the subducting plate melt, which produces volcanoes on the upper plates. The three boundary types also exist in the continents, but here they are more complicated. In particular, subduction under a continent can produce a mountain range as well as volcanoes.

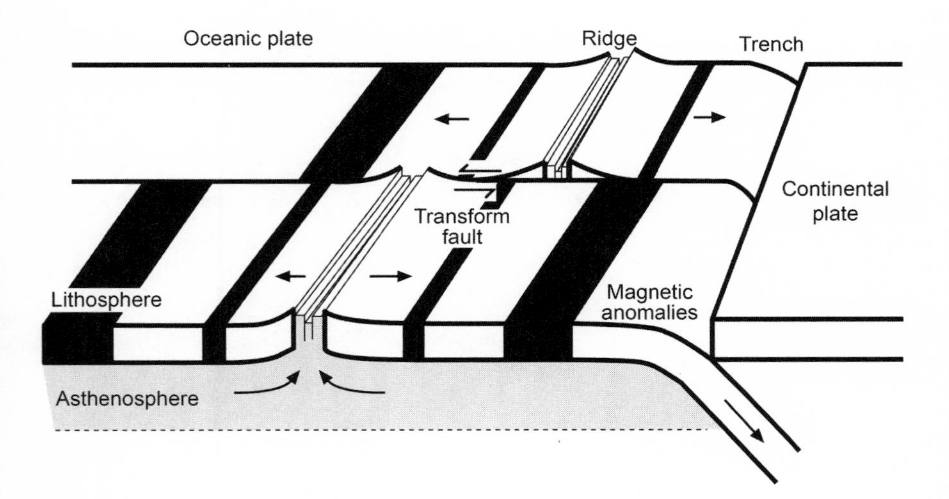

FIGURE 9.3 Types of plate boundaries in oceanic lithosphere. Oceanic lithosphere is formed at ridges and subducted at trenches. At transform faults, plates slide by each other. (Stein and Wysession, 2003)

Plate tectonics explains the toughest problem with Wegener's ideas—what made continents drift. To see this, think of heating a pot of water on a stove. As the water on the bottom gets hotter, it expands, becomes less dense, and rises to the top. Once it gets there, it cools, becomes denser, and sinks again. This process of hot fluid rising and cold fluid sinking is called convection. Plate tectonics is a more complicated version of this simple convection system (fig. 9.4). Mid-ocean ridges are upwelling areas where hot material rises from the deep mantle and cools to form cold, strong plates. Subduction zones are downwelling areas where plates are consumed as their cold material sinks, heats up, and is mixed back into the mantle.

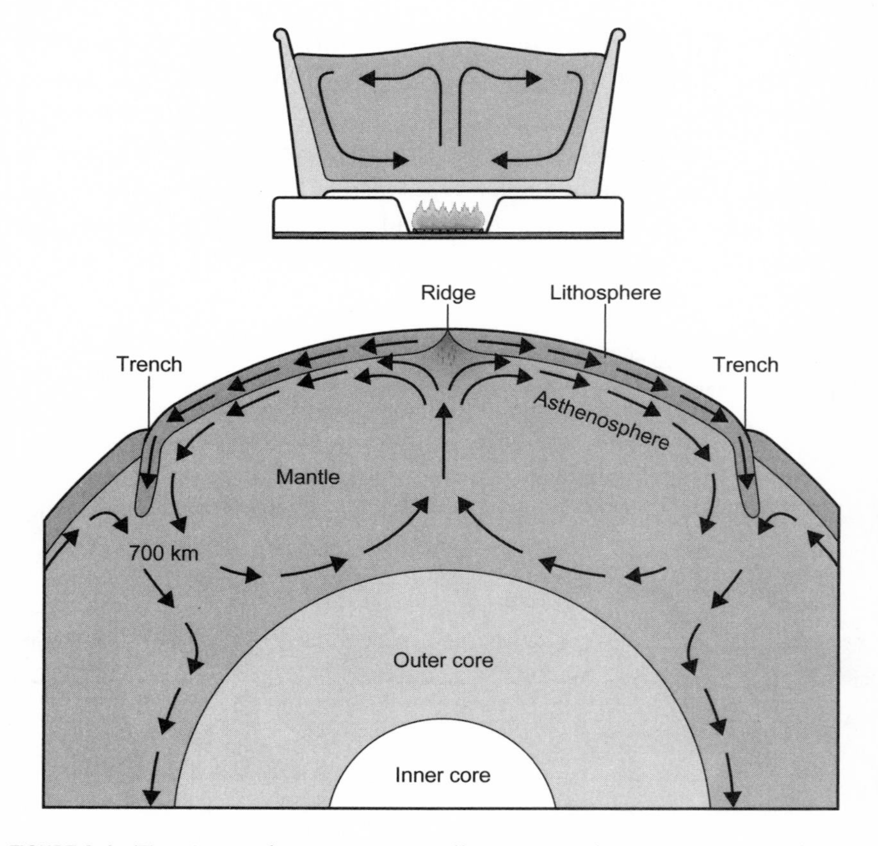

FIGURE 9.4 *Top,* A simple convection cell; *Bottom,* Plate tectonics results from convection in the earth's mantle. The real pattern is more complicated than this simple picture.

The plates that people live on are the cold outer layer—the scum—on top of the convecting mantle. Although we can't directly see how the mantle flows, seismic images and computer models are starting to give us some ideas. These show that the real pattern of three-dimensional flow in the mantle is a lot more complicated than the simple picture. We'll see that this flow might be part of what causes New Madrid earthquakes.

Mantle convection happens because the earth is a giant heat engine. A heat engine is a device that uses heat to make something move. For example, cars run by using the heat from burning gasoline. The earth uses heat rising from its deep interior to move plates. Some of this heat comes from the decay of radioactive elements, and some is left over from heat released when the material in the early solar system crashed together to form the earth 4.6 billion years ago. The planet is almost like a living being whose geological lifeblood is heat. The earthquakes, volcanoes, and mountain building caused by plate motions show that the earth is geologically alive and still evolving.

Realizing that earthquakes result from the planet evolving gives an important insight. People think of earthquakes in terms of the damage they do. However, earthquakes are part of the plate tectonic process that humans need to survive. Plate tectonics makes the continents so humans can live above sea level. Without this, dolphins might rule the world. Gasses produced by volcanoes at plate boundaries maintain the atmosphere that people breathe and keep the earth warm enough for us to live. Plate tectonics also produces the natural resources like oil and gas that power modern society. Thus, earthquakes are part of the price for a planet people can live on.

Where Earthquakes Happen

Plate tectonics explains a lot about earthquakes, starting with why they happen where they do. Figure 9.5 shows the major plates, their boundaries, and the distribution of earthquakes. The plate names are mostly easy to remember: North American, Pacific, and so on. Most earthquakes occur at plate boundaries and so are called "interplate" earthquakes. These earthquakes show that most of the motion occurs at the boundaries between the plates rather than within them.

There are also some earthquakes within plates, called "intraplate" earthquakes. These are a lot less common. For example, the ones at New Madrid

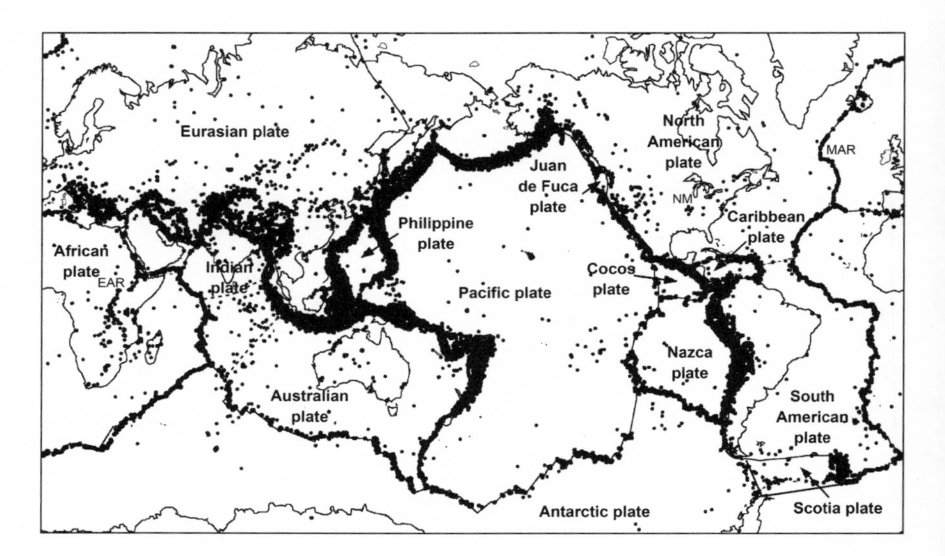

FIGURE 9.5 Map of major plates and earthquake locations, shown by dots. The earthquakes outline most plate boundaries. NM marks New Madrid. MAR is the Mid-Atlantic ridge. EAR marks the East African rift. (Stein and Wysession, 2003)

discussed in this book barely show up on the map. For that reason, we'll talk about interplate earthquakes first.

The locations of earthquakes are one of the ways of figuring out where plate boundaries are. Geologists also use the topography and geology, including the locations of volcanoes.

When the location of a boundary is known, the question is which way the plates on either side move. Until a few years ago, the only way to find out was to combine three different types of data from different boundaries. The speed of plate movements was measured from the rates of spreading at mid-ocean ridges. As the newly formed rock cools, it is magnetized by the earth's magnetic field, so rocks near a spreading center have acquired the earth's present magnetic field. However, rock formed at an earlier time when the earth's magnetic field was reversed acquired the opposite magnetization. As a result, the sea floor contains stripes of rock with opposite magnetizations, so it's like a magnetic tape recorder. Using the paleomagnetic time scale, the stripes show how fast plates move.

Just as the magnetic data gave the speed of plate motion, two other types gave the direction of plate movement. One was the direction of transform faults, because plates slide along them. The other was from the direction of slip in earthquakes at plate boundaries.

Combining data from around the world yields a global plate motion model that describes the motion of all the plates relative to their neighbors. Figure 9.6 shows one that a group of us developed. We called it NUVEL-1 (Northwestern University Velocity) because it was a new ("nouvelle") model. Because this big project took 11 years, some people thought "OLDVEL" might be a better name. Richard Gordon, who is now at Rice University, and I started the project secretly because we weren't sure how authors of the existing models would react. We first collected the data used in the previous model and had it punched onto paper cards, which was how we got data into computers in those long-gone days. Next, we and students compiled and analyzed newer

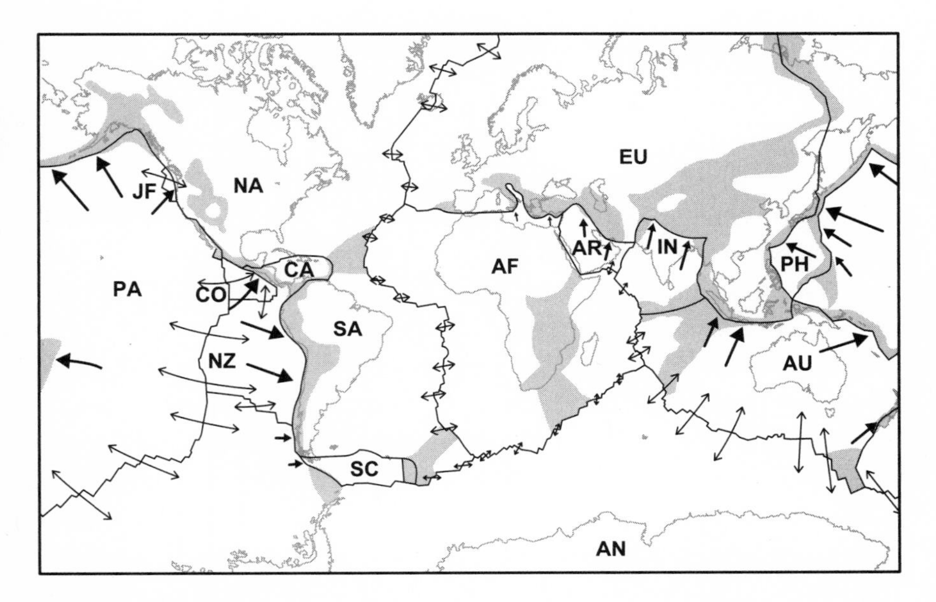

FIGURE 9.6 Major plates and the relative motion at their boundaries. Arrow lengths show relative speed. Diverging arrows show spreading at mid-ocean ridges. Single arrows on a subducting plate show convergence. Stippled areas are diffuse plate boundary zones. (After Gordon and Stein, 1992)

data from around the world. When I told Bernard Minster, an author of the earlier model, what we'd done, he laughed and said he'd have given us their data cheerfully. After finishing that model, he had no desire to do another. This was a nice example of how scientists hand off projects to new people who take the next steps. As the Japanese say, only a fool climbs Mount Fuji twice. Other researchers have since made better models using the new GPS data.

In making these models, we use the mechanisms of earthquakes to study how plates move. We can also go the other way and use what we know about how plates move to predict the focal mechanisms of future earthquakes.

For example, the Mid-Atlantic ridge separates the eastern edge of the North American plate, which contains most of the U.S., from the Eurasian plate. The ridge is a spreading center marked by an underwater mountain chain that shows up beautifully on a topographic map. In addition, the ridge shows up as a band of earthquakes through the middle of the Atlantic.

The west edge of the North American plate is more complicated. Three plates interact along the west coast of North America (fig. 9.7). In the south, the Pacific and North American plates are separating at a mid-ocean ridge that runs down the Gulf of California, the body of water between the peninsula of Baja California and the rest of Mexico. The new lithosphere formed at the ridge widens the gulf and moves Baja California away from the rest of Mexico.

Farther north, the Pacific plate moves northwest relative to North America along the San Andreas fault and related faults. Because the San Andreas fault system is essentially parallel to the relative motion between the North American and Pacific plates, it's largely a transform fault. Even farther north, there's a small third plate called the Juan de Fuca plate sandwiched between the other two plates. The Juan de Fuca plate separates from the Pacific along a spreading center on its west side called the Juan de Fuca ridge. On its east side, it subducts under the North American plate at the Cascadia subduction zone.

The earthquakes and geology reflect these motions. The 1906 San Francisco earthquake resulted from right-lateral strike-slip motion as the Pacific plate on the west side of the San Andreas fault slides northwest relative to the North American plate. The geologists of 1906 had the motion right but thought it occurred only close to the fault. Now, we realize that the motion involves two plates thousands of miles wide.

Farther north, the plate west of the North American plate is no longer the Pacific, so the San Andreas fault ends. Instead, the plate to the west is the Juan de Fuca, so the motion on the plate boundary is different from that on

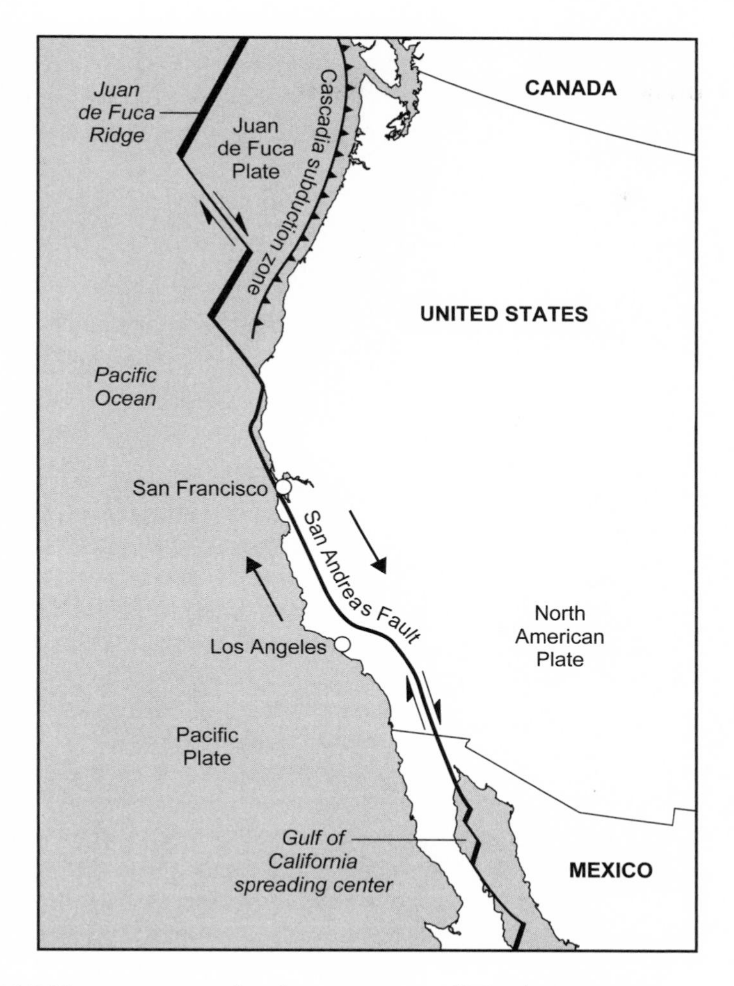

FIGURE 9.7 Plate geometry for the west coast of North America.

the San Andreas. The Juan de Fuca plate subducts beneath North America and produces the Cascade mountain chain of volcanoes in northern California, Oregon, Washington, and British Columbia. The volcanoes include Mount Rainier, near Seattle, and Mount St. Helens, which erupted in 1980 and caused $2 billion in damage—mostly to lumber in the forest—and 57 deaths. Earthquakes also occur on this boundary, including a huge one in 1700 that generated a tsunami that reached Japan. This earthquake was probably a lot like the 2004 Sumatra one, so a giant tsunami is likely the next time such an earthquake happens.

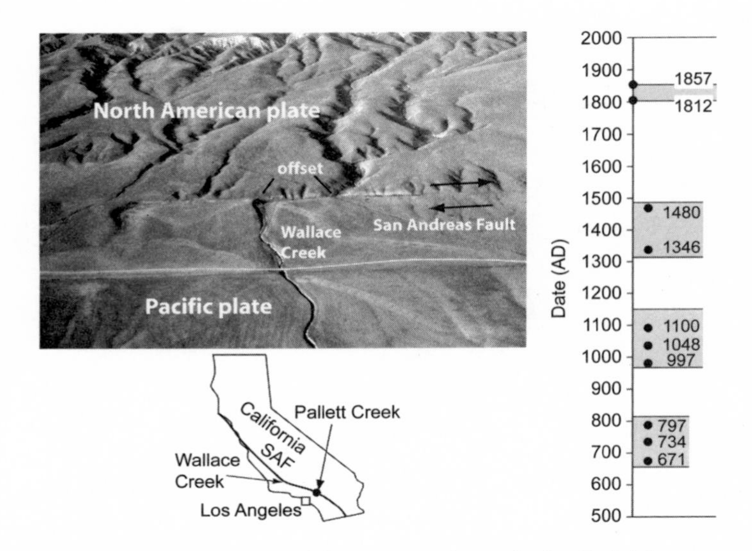

FIGURE 9.8 Geological studies of the San Andreas fault. *Left,* The rate of motion across the fault, and thus between the Pacific and North American plates, is measured using the offset of Wallace Creek (USGS); *Right,* Paleoearthquake history at Pallett Creek. (Stein and Wysession, 2003)

Plate tectonics also explains where the biggest earthquakes are. Subducting plates dip into the earth at a shallow angle, so a large piece is close enough to the earth's surface to have rocks cool enough for stick-slip faulting. Thus, thrust fault earthquakes at subduction zones have the largest fault areas and highest magnitudes of any earthquakes. The magnitude 9.4 Alaskan earthquake of 1964 occurred where the Pacific plate subducts beneath North America, and the 2004 Sumatra earthquake happened where the Indian plate subducts beneath the Burma plate.

When Earthquakes Happen

Plate tectonics also tells us about how often earthquakes happen. A great example is in central California where the San Andreas fault shows up beautifully as a scar across the landscape. It's so clear that geologists joke that pilots flying from Los Angeles to San Francisco follow the fault rather than bother with instruments.

A stream called Wallace Creek crosses the San Andreas south of the part of the fault that broke in 1906 (fig. 9.8). In the past, the creek ran straight across the fault. But as the Pacific plate moved to the left relative to the North American plate, it shifted the creek so that the parts on either side of the fault are now about 425 feet apart. Radioactive dating of charcoal in the streambed shows that this happened during the past 3,700 years. Thus, over this time the average speed of plate motion here is 425 feet / 3,700 years, or 1.4 inches (36 millimeters) per year.

This measurement is a great example of how field geologists use simple methods to solve complicated problems. At the time, there wasn't any other way to measure this motion. In later years, GPS measurements across the fault, taken using the most sophisticated technology in geology, gave the same answer. This site is now a landmark and is preserved in Carrizo Plain National Monument. The stream was named after Bob Wallace, one of Northwestern's most famous geology graduates, who made the measurements. In his words, "The geology department at Northwestern had, and as I understand it, still has, a sense of family. The professors paid close attention to the students and took personal interest in them." Hopefully, we can continue to inspire scientists like him.

After seeing the beautiful San Andreas, it's disappointing that faults in the New Madrid area don't show up this way in the landscape. Part of the reason is that the New Madrid ones are buried, and part is that the area gets a lot more rain, which erodes topography. Still, the difference tells us something important. Over time, there's been a lot less motion across the New Madrid faults.

Measuring the rate of motion across the San Andreas gives us a way to estimate about how often big earthquakes happen. Although the plates move at this speed, elastic rebound theory says that most years the fault doesn't slip at all. As the plates on either side move, the fault stays locked until a big earthquake. In 1906, the fault slipped about 13 feet, or 156 inches. Building up 156 inches at 1.4 inches per year would take 111 years (156/1.4). So if the big earthquakes on this part are like the 1906 one, they should be about 110 years apart.

That seems about right. Geologist Kerry Sieh compiled a beautiful record of past earthquakes at a nearby site on the San Andreas called Pallett Creek using sand layers that were disturbed by earthquake shaking. These kinds of studies are called paleoseismology. The most recent large earthquake on this part of the fault was in 1857, when historical records tell us that an earthquake about the size of the 1906 one happened. As shown in figure 9.8, it looks like earlier large earthquakes occurred approximately in the years 1812, 1480, 1346, 1100, 1048, 997, 797, 734, and 671. The average time between the paleoearthquakes is 132 years, which is about what the stream offset shows.

The fact that plate tectonics tells a lot about how often big earthquakes happen is very useful. After the tsunami from the 2004 Sumatra earthquake devastated the coast of Thailand, a natural question was how soon a similar tsunami would happen again. At the time, there wasn't a history of tsunamis there. Using the slip in the earthquake and plate motions, my colleague Emile Okal and I calculated that such earthquakes were at least 400 years apart. Thus, instead of abandoning damaged beach resorts, it made sense to rebuild them. We didn't have the money to buy the resorts, which would have been cheap because people worried that a similar tsunami would strike soon, so we settled for a paper in a scientific journal. Geologists have since found deposits showing that the last tsunami was about 600 years ago.

It would be nice to use these methods for New Madrid, but there's a problem. New Madrid is in the middle of the North American plate. If plates behave perfectly according to plate tectonic theory, there shouldn't be any movement there at all. That's not much help.

The Pallett Creek earthquake history makes a nice problem for students, whom I ask to predict when the next earthquake will happen. Because the average time between earthquakes is 132 years, their first guess is 1989, when it didn't happen. Asked why, they realize that the earthquake history is more complex than we'd think just from the average time between them. The intervals between earthquakes vary from 45 years to 332 years, and it looks like something more complicated is going on. The earthquakes seem to come in clusters: three earthquakes between 671 and 797, a 200-year gap, three between 997 and 1100, and then a 246-year gap. Hence, using the earthquake history to forecast the next big earthquake is complicated. If the recent cluster is still going on, an earthquake might happen soon. On the other hand, if the cluster that included the 1812 and 1857 earthquakes is over, it might be a long time until the next big earthquake.

There are other complexities. The dates have uncertainties. Some earthquakes might have been missed, and some earthquakes might have been bigger than others. All of this shows that although these methods give an average time between earthquakes, they can't predict the next one very well. Scientists have tried to give probabilities of how likely an earthquake is, but these aren't that useful. Studies can produce a wide range of numbers, depending on how much of the history is used and what's assumed. For example, a study in 1989 estimated that the probability of a major earthquake before 2019 was somewhere between 7 and 51%. As we'll see, all these problems are even worse when researchers try to estimate how likely a large New Madrid earthquake is in the next few hundred years.

Even so, plate tectonics gives a good idea of why the level of earthquake activity varies from place to place. Because earthquakes result from plate motion, Chile—where the Nazca plate subducts beneath South America at a speed of about 80 millimeters per year—should be about twice as active as California where the plate motion is about 40 mm/yr. Similarly, California should be about six times more active than Haiti, where the plate motion is about 7 millimeters per year. This is in turn faster than the 2 millimeters per year across Utah's Wasatch fault. All of these are more active than New Madrid, which is within the North American plate and so deforms at much less than 1 millimeter per year.

Earthquakes Near Plate Boundaries

When I started graduate school in California in 1975, the relationship between earthquakes and plate tectonics had been discovered only a few years earlier. My fellow students and I went to the San Andreas fault, put one foot on either side, and thought we had one foot on the Pacific plate and one on the North American plate.

We should have had more sense. All of us knew that large earthquakes occur in California on faults other than the San Andreas. In fact, they happen across the western U.S., including in Nevada, Utah, and Montana. We knew that some of these could be almost as big as the largest ones on the San Andreas. For example, in 1959 a magnitude 7.3 earthquake near Hebgen Lake, Montana, killed 28 people and triggered a huge landslide that dammed the Madison River. Still, we didn't have the sense to wonder how these could happen east of the San Andreas, which geologists at the time thought of as "the" plate boundary.

A few years later, geologists started thinking about this question. It turned out that the rate of spreading between the Pacific and North American plates measured from magnetic stripes in the Gulf of California was 48 millimeters (1.9 inches) per year, or about 35% faster than the 36 millimeters (1.4 inches) per year across the San Andreas at Wallace Creek. This difference doesn't sound like much, but geology deals with long times. In a million years, movement at the rate of a half-inch per year gives a distance of seven miles. The difference showed that although most of the motion between the plates occurs across the San Andreas, the extra motion is spread out across a boundary zone that spans the western third of the U.S. Once GPS came along, it showed this extra motion beautifully. The motion produces spectacular mountains and valleys and is why Salt Lake City has an earthquake hazard although it's almost a thousand miles from the San Andreas.

A funny thing happened when sorting this out. Before we began looking into this at Northwestern, the spreading rate in the Gulf of California was thought to be 58 millimeters (2.3 inches) per year. This meant that even more of the plate motion was taking place off the San Andreas. One possibility was that some took place west of the San Andreas, off the coast of California. This was a hot topic because motion on offshore faults could cause an earthquake hazard for nearby nuclear power plants. When graduate student Charles DeMets, now a professor at the University of Wisconsin, showed me a new analysis of the magnetic stripes giving the lower rate, I pointed out that it greatly reduced the perceived earthquake hazard. Within a few years, GPS measurements showed that we were right. It was nice to have resolved a question about geology in California, where many leading earthquake scientists live, without leaving Chicago.

Because similar situations occur along other plate boundaries, we now talk about plate boundary zones rather than narrow plate boundaries. The earthquake map (fig. 9.5) shows many earthquakes that we used to think were inside plates but now realize are in a wide boundary zone. For example, a broad band of earthquakes marks the boundary zone along which the Indian plate is colliding with the Eurasian plate, which has built the Himalaya Mountains. There's also a zone of earthquakes along the East African Rift, where Africa is splitting up into two plates. Still, some earthquakes, like those at New Madrid, are in the interior of plates, nowhere near a boundary zone.

This change from thinking about narrow plate boundaries to broad plate boundary zones shows what it means to talk about the "theory" of plate tectonics. In daily life, a theory is a vague idea with little to back it up. However, in science, a theory is a model of how the physical world works. It's based on observations that can be used to predict what will happen and can be tested using new data. A scientific theory describes a wide range of phenomena, even if it isn't a full description. For example, Newton's theory of gravity tells us how objects fall. It's a great theory because it matches experiments beautifully and is used to navigate spacecraft to other planets. Still, it's not a complete theory because it doesn't tell what causes gravity, which can be described only by relativity. Thus, as needed, theories get changed.

In our case, plate tectonics explains a lot about how the earth works, including why, where, and how often most earthquakes occur. The answers are clear for plate boundaries, where the earthquakes are a direct result of the plate motions. However, everything is more—you guessed it—complicated for earthquakes inside plates, like the ones at New Madrid. These earthquakes are an indirect result of plate tectonics, as we'll see next.

Chapter 10

Earthquakes That Shouldn't Happen

Anything that did happen, can happen.

—Geologist Marshall Kaye, 1904–1975

We've just seen that almost all of the world's earthquakes occur along the boundaries between plates. That's because earthquakes happen when forces in the earth make faults move. Plate tectonics explains that many faults are part of the boundary between plates and that the force to move them comes from the motion between the plates on either side. Earthquakes inside plates are rarer because the movements between plates are much faster than motion within plates. If plates behaved perfectly there'd be no motion within them, and earthquakes wouldn't happen there. Places like New Madrid wouldn't have earthquakes.

Intraplate Earthquakes in North America

That's almost true. As the top map in figure 10.1 shows, almost all earthquakes in North America are in the western plate boundary zone. There are some intraplate earthquakes in the central and eastern part of the continent, but not many. Because the map shows earthquakes since 1900 plus some older earthquakes, we have to plot down to magnitude 5 earthquakes—which aren't very big and rarely do much harm—to clearly see the New Madrid zone.

These earthquakes show that plates aren't perfectly rigid. There's some motion inside them. What causes these small motions and the earthquakes that result is an interesting problem. Still, most seismological research studies the bigger and much more common plate boundary earthquakes, and I've

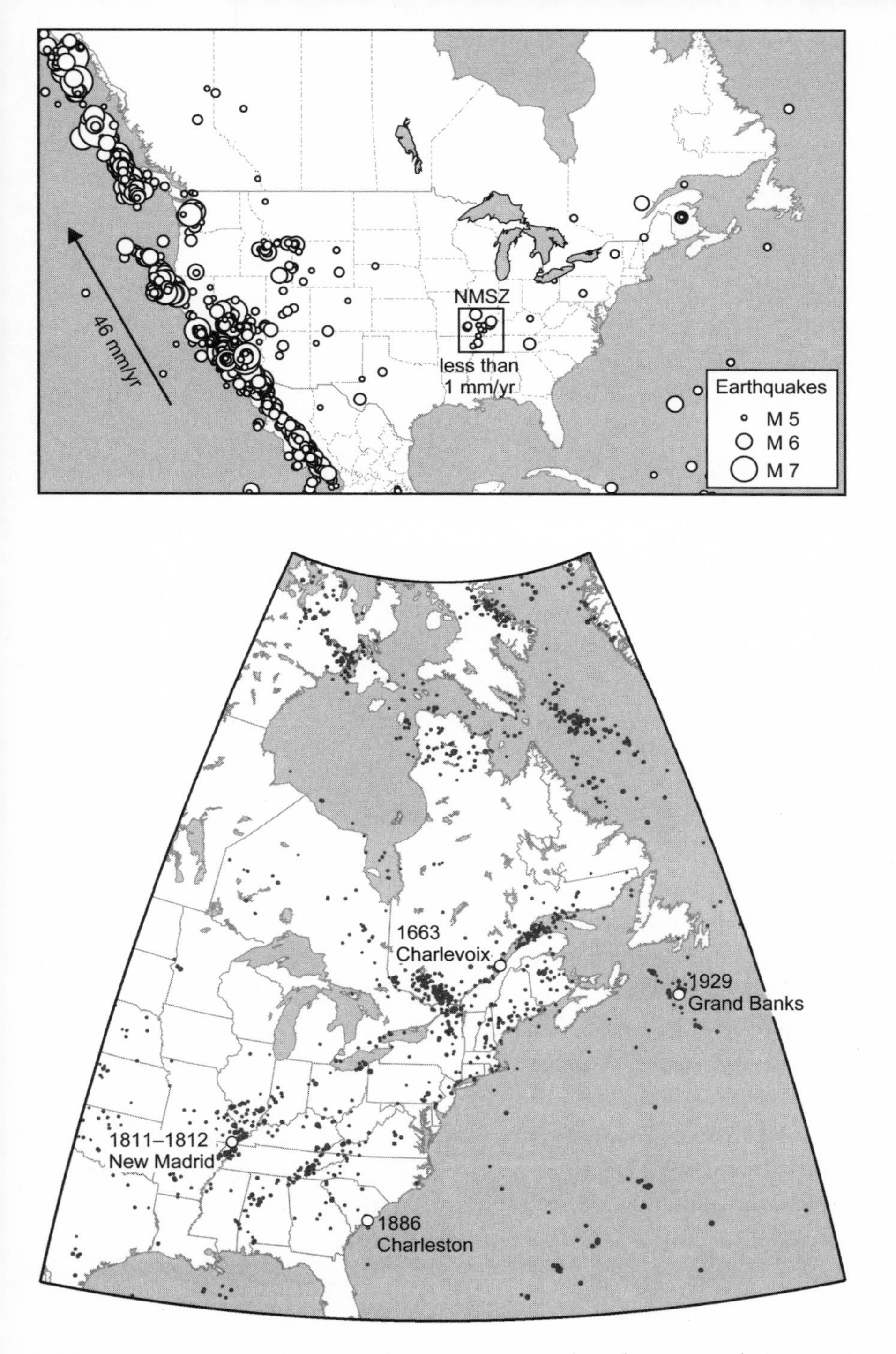

FIGURE 10.1 *Top,* Map of magnitude 5 or greater earthquakes in North America (Stein et al., 2003); *Bottom,* Map of magnitude 3 or greater earthquakes in eastern North America since 1973. (Mazzotti, 2007)

been asked whether the ones in continents are worth bothering with. A sensible answer might be "no," but I've been interested in them for years because they don't fit easily into plate tectonics.

Studying earthquakes within continents shows how some of the ways geologists tackle a problem are different from what other scientists do. Physicists and chemists study as simple a system as they can, usually in a lab. Geologists deal with long time periods and a planet that can't be isolated in a lab. One of our heroes, the famous environmentalist John Muir, said "When we try to pick out anything by itself, we find it hitched to everything else in the universe." That's why geologists tend to look at a big picture. One method is to find examples of what we're studying in other places and times and look for patterns that will give us some ideas about what's going on.

Let's begin by looking at the earthquakes inside North America, away from the plate's boundaries. In geology, a good way to start is often by looking for patterns on a map. The bottom part of figure 10.1 zooms in on earthquakes in eastern North America by leaving out all the action in the west and plotting down to magnitude 3. This shows a lot more earthquakes than the upper map because there are about 100 times more magnitude 3s than 5s. The little earthquakes make the map look scary but aren't a problem because a magnitude 3 releases 1,000 times less energy than a 5. The map also shows all the known large earthquakes—magnitude 6 or above—since the 1600s, when we first have historical records. There aren't many of these earthquakes, but there are enough to show some patterns.

New Madrid shows up nicely on the map. We see the big earthquakes from 1811 and 1812 and the smaller earthquakes since then that map out the seismic zone. There are also other earthquake zones that are active by eastern North America standards, even if they wouldn't impress a Californian. Several of these have had earthquakes about as big as the 1811–1812 New Madrid ones, based on the fact that they had similar patterns of shaking.

The best analogy is the earthquake zone along the Saint Lawrence River valley in Canada. A big earthquake, perhaps around magnitude 7, happened in 1663 near the town of Charlevoix. It caused major landslides along the river and was felt in Massachusetts, where it did minor damage. There have been smaller earthquakes here since.

A common place for both large and small earthquakes is along the east coast. One with a magnitude of about 7 struck near Charleston, South Carolina, in 1886. From the reported intensities of shaking, the earthquake seems to have been nearly as big as the largest New Madrid earthquakes. Descriptions of what happened are similar, including the formation of sand features. The earthquake destroyed or damaged many buildings and killed about 60 people. As in San Francisco 20 years later, much of the city was on filled land, and many buildings were built of brick, which is easily damaged by earthquake shaking. It took the city about five years to recover from the earthquake, and some of the damaged buildings can be seen today. These include the College of Charleston's beautiful Randolph Hall (fig. 10.2), which appeared in the film *The Patriot*. What happened is useful for thinking about the effects of a possible large earthquake in the New Madrid zone, where brick construction on soft sediment is common.

Similar-size earthquakes happened elsewhere along the coast. In 1933, a magnitude 7.3 earthquake happened in Baffin Bay, and in 1929, a magnitude 7.2 earthquake occurred off the Grand Banks of Newfoundland. Shaking from the Grand Banks earthquake was felt as far south as New York but did only minor damage on land. However, the earthquake generated a huge underwater landslide that moved about 50 cubic miles of sediment, causing a tsunami that destroyed homes, ships, and businesses. The 27 people killed by the tsunami are the only known deaths in Canada from an earthquake to date, though this might change when a major earthquake strikes the Cascadia subduction zone on the west coast.

FIGURE 10.2 *Left,* Randolph Hall with repairs after the 1886 earthquake that appear as lines from the corners of the windows and as the metal "earthquake bolts" under the windows (Erin Beutel); *Right,* Sand crater formed by the earthquake. (USGS)

The Grand Banks earthquake gives us insight into the magnitudes of the big New Madrid earthquakes of 1811 and 1812. The pattern of the intensity of shaking versus distance from the epicenter is like those for the New Madrid ones. Because the seismic waves from both traveled through the rocks of eastern North America, these shaking patterns show that the earthquakes were about the same size. Seismograms from the Grand Banks earthquake show that it had a magnitude of about 7.2. That's part of the reason for thinking that the 1811 and 1812 earthquakes had magnitudes less than 7.5. Magnitude 8 earthquakes would have shaken much larger areas.

A smaller earthquake, with a magnitude of about 6, struck near Cape Ann, north of Boston, in 1755. Many buildings, especially those on filled land near the harbor, suffered minor damage, mostly to chimneys. "It was a terrible night," wrote the Reverend Mather Byles, "the most so, perhaps, that ever New England saw." Smaller earthquakes have occurred nearby since.

White Rocks and Black Rocks

These examples show that earthquakes in the Midwest are part of a pattern of earthquakes in the continent. What is it about the continents that lets these earthquakes happen even without a plate boundary?

The answer is white rocks and black rocks. To see why, let's ask the question: Why are there continents? Put another way, why is about 30% of the earth's surface above sea level while the rest is covered by the oceans? The difference is huge. Oceans are really deep, with an average depth of about 12,000 feet below sea level. Continents are on average about 2,400 feet above sea level. This might be hard to believe in the Midwest, but it's true. New Madrid is about 300 feet above sea level, and there's a lot of higher land to the west.

These different heights happen because the rocks of the earth's crust under continents are different from those under the oceans. Crust under the continents is pretty much like granite, the white rock in figure 10.3. Granite occurs in places like Missouri's Saint Francis Mountains, California's Sierra Nevada, and New Hampshire's White Mountains. When these rocks are eroded by rain, wind, and ice, the sediments that result are carried by rivers and end up in places like beaches, giving the beautiful white sand along the

FIGURE 10.3 Samples of granite (*left*) and basalt (*right*). (Kim Adams)

Great Lakes' shores. Some of this sand turns into the sandstone rock that appears in many places in the Midwest.

In contrast, the crust under the oceans is mostly basalt, the darker rock in the picture. Basalt is the volcanic rock that forms plates at mid-ocean ridges. It's not as common on the continents as granite, but there's some. It appears in cliffs along the St. Croix River between Wisconsin and Minnesota and in volcanic areas like Yellowstone National Park.

Granite and basalt have different colors because they have different chemistry. Granite contains mostly the elements silicon and oxygen, while basalt has less of these and more iron and magnesium. That difference makes granite about 15% less dense than basalt, which means that a chunk of granite weighs about 15% less than a chunk of basalt the same size.

This density difference has a huge consequence for how the earth works. Because the continents are made up of less dense granite, they are higher than the denser basalt rocks under the oceans. The continents "float" above the denser basalt, just the way wood floats in water because it's less dense. For the same reason, the rocks forming the continents don't sink into the mantle at subduction zones. They rose to the surface early in the earth's history. Since then, although they have been battered and reshuffled by plate tectonics, they haven't been subducted. This is part of the reason why there are earthquakes at New Madrid.

How Continents Are Injured but Survive

As usual, Alfred Wegener had it right. In his idea, the continents were together 300 million years ago and drifted apart. All the continents are still here, just arranged differently. For example, 300 million years ago India was tucked between Africa and Antarctica. Now it's far from them but crashing into Eurasia.

This works because continents and oceans—specifically the lithosphere under the oceans—have different life histories. Oceanic lithosphere is formed at mid-ocean ridges and destroyed at subduction zones, but continental lithosphere lasts.

This life cycle, one of the basic themes in geology, explains how plate tectonics works over hundreds of millions of years. It's called the Wilson cycle after Canadian geologist J. Tuzo Wilson. Tuzo, as people called him, was a brilliant creative thinker. He first argued that the earth was contracting, then that it was expanding, and finally for plate tectonics. He was great at posing ideas, figuring out how to test them, and dropping ones that didn't work. In addition to this cycle, Tuzo discovered transform faults.

To understand the cycle, geologists use one of our most powerful methods—seeing how the past worked by looking at places where the same thing is happening today. We say that the present is the key to the past. That's nice because it's a lot easier to see the present. For example, if we didn't know how people grew up, we could get a good idea by looking around and seeing babies, toddlers, children, teenagers, young adults, and older adults.

Figure 10.4 shows the stages in the Wilson cycle. It begins when part of a continent starts to be pulled apart. This involves heating from below, but we still don't know exactly how. The granite crust stretches like taffy and starts to break along newly formed faults, causing earthquakes and forming what's called a rift valley.

Sometimes, rifting goes on for a while and then stops. For unknown reasons, it fails to split the continent. Instead, it leaves a "failed rift," a long valley of stretched and faulted rock that eventually gets filled up and buried by sediments. Failed rifts appear in continents all over the world, as pointed out by Kevin Burke and John Dewey, who had a major role in explaining how continental geology reflects plate tectonics. I met Kevin when I was in graduate school and admired both his insight and ability to draw a geologic map of almost any part of the world from memory.

The failed rift idea explains what happened at New Madrid starting about 750 million years ago. Researchers think the earthquakes happen today, deep below the sediments laid down by the Mississippi River, because the old

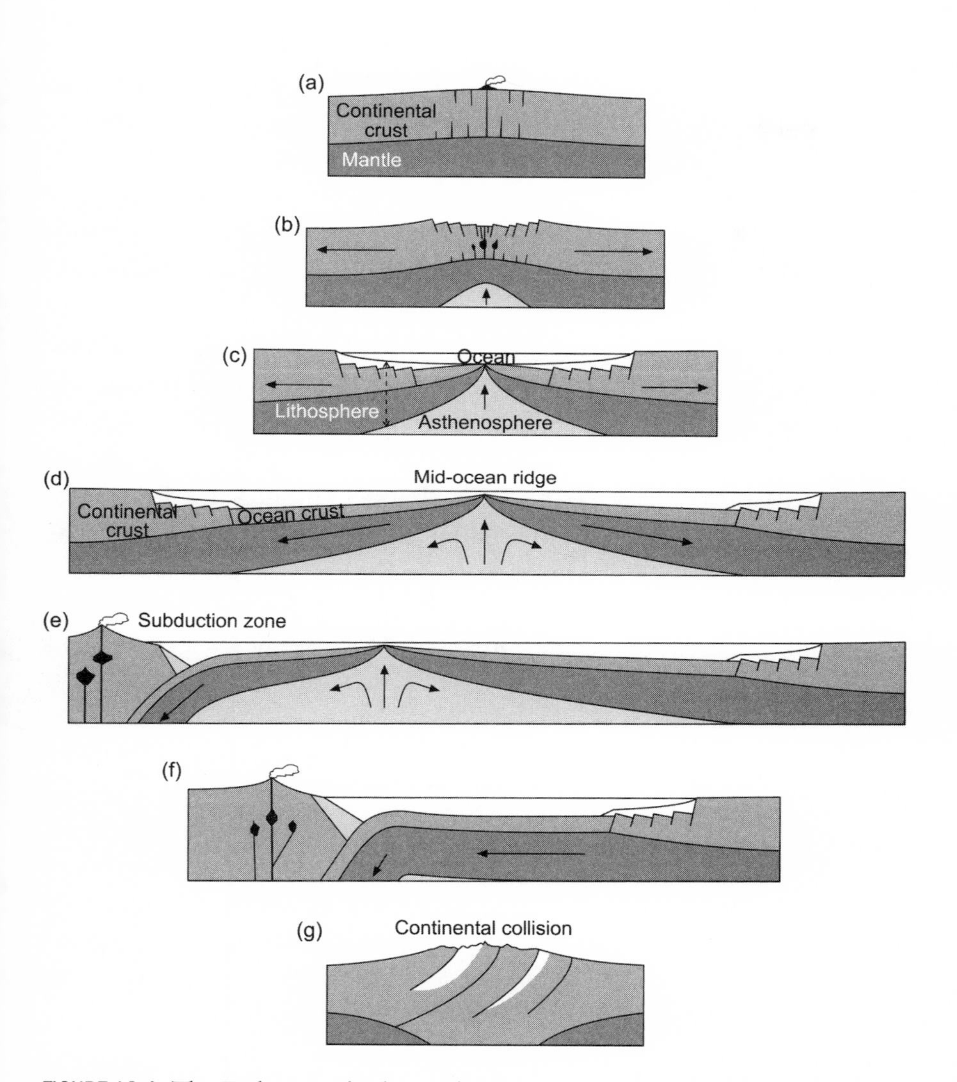

FIGURE 10.4 The Wilson cycle shows the stages through which the continents and oceans evolve. *(a), (b),* Continental stretching and rifting; *(c)* Seafloor spreading begins, forming a new ocean basin; *(d)* The ocean widens and is flanked by sedimented passive margins; *(e)* Subduction of oceanic lithosphere begins on one of the passive margins, closing the ocean basin; *(f)* and starting continental mountain building; *(g)* The ocean basin is destroyed by a continental collision. (Stein and Wysession, 2003)

faults can still move if something pushes them. Old faults are like broken bones: They heal but not perfectly.

If the rift keeps opening, hot material from the mantle rises under the rift and causes volcanoes where basalt magma erupts. This is happening today along the great East African rift that is splitting Africa in two. Eventually, the rift is filled by enough basalt that it becomes an oceanic spreading center. Spreading at the new ridge forms a new ocean that separates the continental rock on both sides. These young oceans, like the Gulf of California, are narrow because there are only a few million years worth of new crust there.

With time, the ocean widens and looks like the Atlantic does today. The older seafloor, away from the ridge, cools off. The cooler rock contracts and gets denser, so the ocean gets deeper the farther it is from the ridge. Sediments produced when the continents on either side erode accumulate in thick piles along the coast. These sediment accumulations are important because they contain the remains of small living things like plankton and algae, some of which can eventually be turned into oil and natural gas by high temperatures and pressures. That's why oil and gas are often found near the coast or offshore.

As we've seen, earthquakes happen along these coasts, like the east coast of North America. These coasts are called "passive" continental margins because there's no plate boundary there. Both the continent and ocean are part of the same plate. An "active" continental margin is one like that off Oregon and Washington, where there's a plate boundary so the two sides are on different plates. Even without a plate boundary, passive margins can have large earthquakes near the coast, like the Charleston earthquake, and offshore, including the Grand Banks one. Just like at New Madrid, it seems that faults left over from rifting never fully heal.

The ocean can't keep getting wider forever. Eventually, the plate breaks along one of the passive margins, and a new subduction zone forms. The plate being subducted gets smaller because it subducts at one side faster than it's produced at the ridge on the other side. This is happening now to the Juan de Fuca plate along the west coast.

Eventually, the entire ocean basin becomes closed, so the continents on either side collide. This is happening today. The ocean basin between India and Eurasia has been subducted, and the collision between the two continents is pushing up the Himalayas, the highest mountains on earth. They go up because continents don't subduct. Eventually, the mountain building stops. This leaves two continents plastered together with a mountain range between them. There's no motion between the two anymore, so they're now both on one plate.

If more continents merge, a supercontinent forms. That happened 300 million years ago, when all the continents merged to form the supercontinent of Pangaea. Usually, things don't go as far as forming a supercontinent. Before that happens, rifting starts within a continent, and the cycle starts again. Often, the new rifting starts near the site of the earlier rifting because continents don't heal well.

The eastern U.S. is a nice example. The Appalachian Mountains formed in a continental collision that closed an earlier Atlantic Ocean about 300 million years ago as part of Pangaea's formation. Since then, the present Atlantic Ocean opened during the past 200 million years as Pangaea broke up. There are volcanic basalt rocks in places like the Connecticut River valley and along New York's Hudson River that were produced when the present Atlantic started rifting. At the time, these places looked like today's East African rift.

All this means that continental and oceanic crust have very different life cycles. Ocean crust gets subducted, so it never gets older than about 200 million years. Because the less dense continental crust isn't subducted, the continents are much older. The deep rocks around the New Madrid seismic zone are more than 1.5 billion years old, and other parts of the continent are even older. That means they're full of fossil faults, old weak zones where earthquakes can—and do—happen. Thus, plate tectonics, via the Wilson cycle, explains the first question about earthquakes inside continents, namely where did the faults come from?

New Madrid as a Failed Rift

The New Madrid earthquakes probably occur on very old faults, much older than the ones that cause the earthquakes along the east coast. The east coast ones formed when the continents split apart to form the Atlantic Ocean. The ones at New Madrid come from an earlier time when continents rifted.

About 800 million years ago, the continents were assembled into a supercontinent called Rodinia, which is Russian for "homeland." Most of eastern North America was in a part called Laurentia. Paleomagnetists are still working on exactly how it looked, but it was something like what's shown in figure 10.5. Starting around 750 million years ago, Rodinia rifted apart.

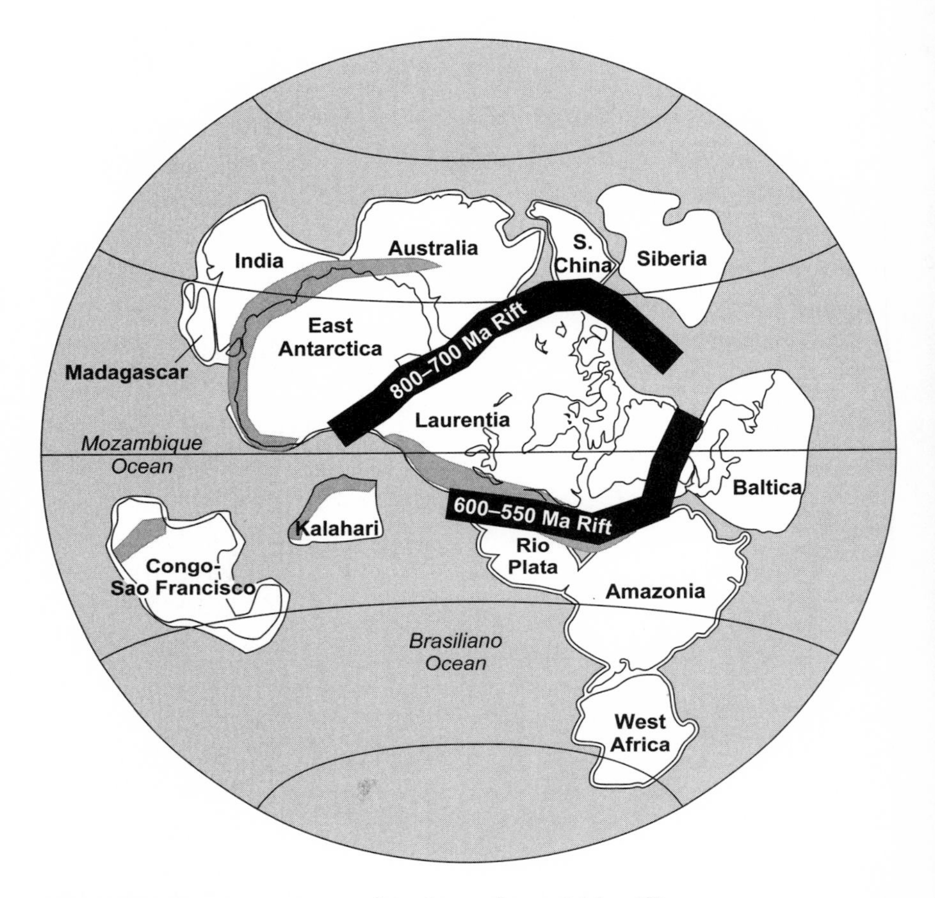

FIGURE 10.5 Reconstruction of Rodinia about 800 million years ago showing where the rifting that isolated Laurentia occurred. (After Meert and Torsvik, 2003)

New ocean basins formed and separated Laurentia from the other assembled continents. The ocean to the east of North America became an early version of the Atlantic. This earlier Atlantic was subducted away when the continents were reassembled into the most recent supercontinent, Pangaea.

For our story, what matters is the part of the rift along the south side of what is now North America called the Reelfoot rift. This rift opened from about 600 to 550 million years ago but died instead of splitting the continent and forming a new ocean. The failed rift left fossil faults that never healed. Since then, the area has had a complicated history. Several major geological events probably made the old faults move. The area was squeezed about 300 million years ago when the continents came together again to form Pangaea. Between about 200 and 90 million years ago, the area was pulled apart and heated, for reasons that are being debated. What happened might have been part of the breakup of Pangaea and the formation of the Gulf of Mexico, or due to heating under the plate by a "hot spot" like the one that's forming the islands of Hawaii.

It's impressive that the faults on which today's New Madrid earthquakes happen formed more than 500 million years ago. Geologists work with times that are so long compared to anything in our daily lives that they're hard to imagine. The earth formed about 4.6 billion years ago and the oldest rocks are about 4 billion years old. The oldest life, simple bacteria, is about 3.5 billion years old. Land plants evolved about 400 million years ago, dinosaurs died about 65 million years ago, and modern humans developed about 100,000 years ago.

However, the faults at New Madrid have moved only occasionally since they formed. In particular, the earthquakes occurring today probably have been going on for only a few thousand years or less. There are several reasons to think this, starting with the fact that if they'd been going on a lot longer, there would be clear topography at the surface.

The idea that today's New Madrid earthquakes occur on faults that formed in a failed rift makes sense, but is only part of the story. It says why the faults are there but not what makes them move in earthquakes. It also doesn't say why there are earthquakes in this particular place, when the continent is full of old faults and weak zones. Still, it's a good start, so we'll pursue it in the next chapters.

Chapter 11

What's Going on Down There?

All models are wrong. Some models are useful.

—George Box, statistics pioneer

The idea that the New Madrid earthquakes happen on fossil faults formed in a failed rift is a "model." In science, a model isn't a small replica, like a model train. It's a fleshing out of an idea into something more specific. Sometimes it's a set of equations, and in other cases, it's a picture. Either way, a model should have enough detail to guide our thinking. It should be "testable," meaning that it predicts things that can be tested with new data to decide whether the model is any good and to change or discard it if it isn't.

Models aren't reality, because they simplify the complicated world. Still, if they simplify it well, they get the important parts mostly right. They're useful if they help us make progress without leading us astray. That's why scientists are always developing, testing, improving, or rejecting models.

The failed rift model—like most models in geology—is based on a mix of data, ideas about how the earth works, guesses, and simplifications. Let's look at what this model says, what data and ideas went into it, what it does and doesn't tell us, and how useful it is.

The Failed Rift Model

In the model, developed by Patrick Ervin and Lyle McGinnis of Northern Illinois University, the granite rocks of the crust were stretched and faulted, leaving a deep valley called the Mississippi Embayment. Over millions of years, the valley filled with thousands of feet of thick sediment deposited by the Mississippi River and its ancestors. The buried Reelfoot rift that got all this started is in the middle of the broader embayment. There are fossil faults beneath the sediment both within the rift and along its edges. These faults never healed, and earthquakes happen on some of them. The big 1811–1812 earthquakes and the smaller ones today that look like aftershocks occur on faults within the rift. Figure 11.1 shows an interpretation of what the area looks like today, developed by a group that included Larry Braile and Bill Hinze of Purdue University and Randy Keller of the University of Texas, El Paso.

In the model, the rift extends northward from New Madrid and splits into three arms. The northeast arm is called the Rough Creek graben ("graben" is the German word geologists use for a steep-sided valley). The northern branch, called the Southern Indiana arm, extends into Illinois and Indiana, where earthquakes occur on faults in the Wabash valley. The northwestern arm is called the St. Louis arm.

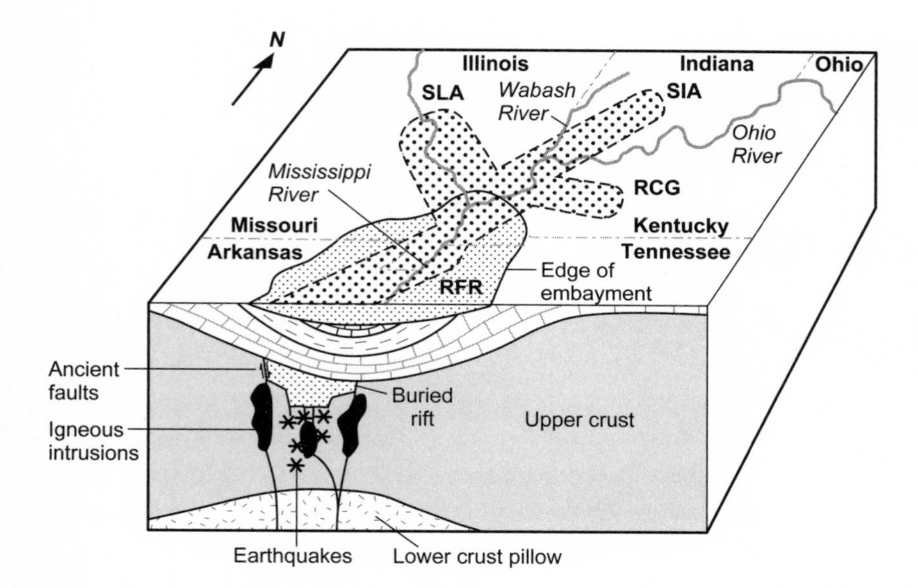

FIGURE 11.1 A geologic model showing how New Madrid earthquakes occur on fossil faults in the Reelfoot rift (RFR). Earthquakes also occur on the rift's extensions: the Rough Creek graben (RCG), the Southern Indiana arm (SIA), and the St. Louis arm (SLA). (After Braile et al., 1986)

Many of the earthquakes that occur north of the New Madrid area are associated with these arms. Seismologists often call the whole area the New Madrid seismic zone, but when they're being specific, they use "New Madrid seismic zone" for the most active part where the big earthquakes of 1811 and 1812 happened, and terms like "Wabash Valley seismic zone" for the other areas. Unlike some other sciences, seismologists don't have committees that define precise terms that must be used. That's in part because of how complicated the earth is and in part because we're too individualistic.

Getting the Picture

I often start my beginning geophysics course by asking students to figure out what's inside a paper "mystery box" containing a heavy weight wrapped in newspaper. Everyone shakes the box and decides that there's a small object inside. Next, they heft the box and decide that the object inside is pretty heavy, perhaps a brick or a piece of metal. When asked how to choose between these options, students suggest trying a magnet. The students have quickly figured out three methods that geophysicists use to look into the earth.

Seismologists shake the earth using seismic waves. To see down to a depth of a few miles, they use waves produced by earthquakes or make waves themselves. Because people don't want big explosions nearby, seismic waves are generated by special trucks with hydraulic weights that vibrate up and down. This system, called Vibroseis, was invented by seismologist Selwyn Sacks in his Ph.D. thesis and patented by his university, but became such a success in the oil industry that the patent was ignored. Seismometers record the waves that bounce off different rock layers (fig. 11.2). Trucks and seismometers move across an area, collecting seismograms.

Putting many seismograms next to each other gives a picture of the earth below, called a "seismic section." The vertical axis is the time when a reflection arrives. The top is time zero and corresponds to the earth's surface. Lines across the picture show when waves arrive at times corresponding to the depth of the layers the waves reflected from.

Figure 11.3 shows a seismic section across part of the Southern Indiana rift arm from a survey by John Sexton of Southern Illinois University and others. Combining the seismic section with what's known about the geology

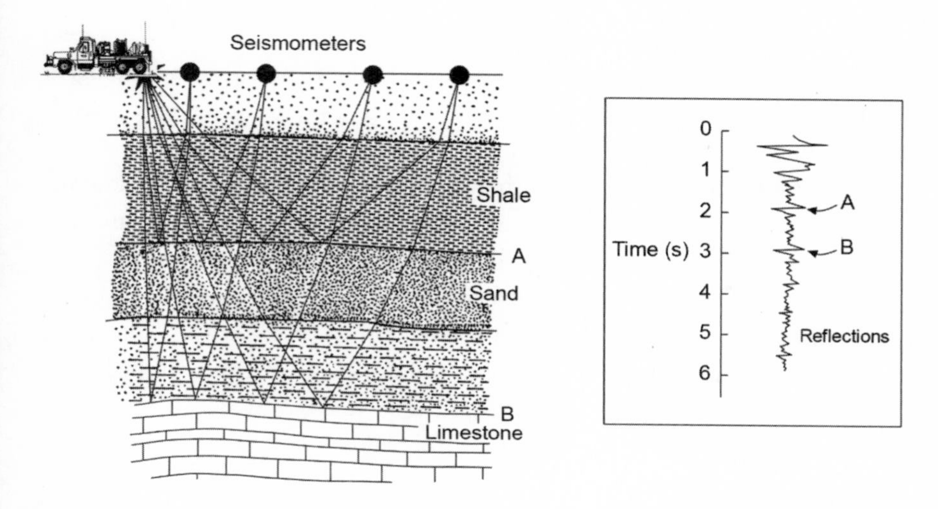

FIGURE 11.2 Seismic waves are used to map the geology at shallow depths in the earth. (Stein and Wysession, 2003)

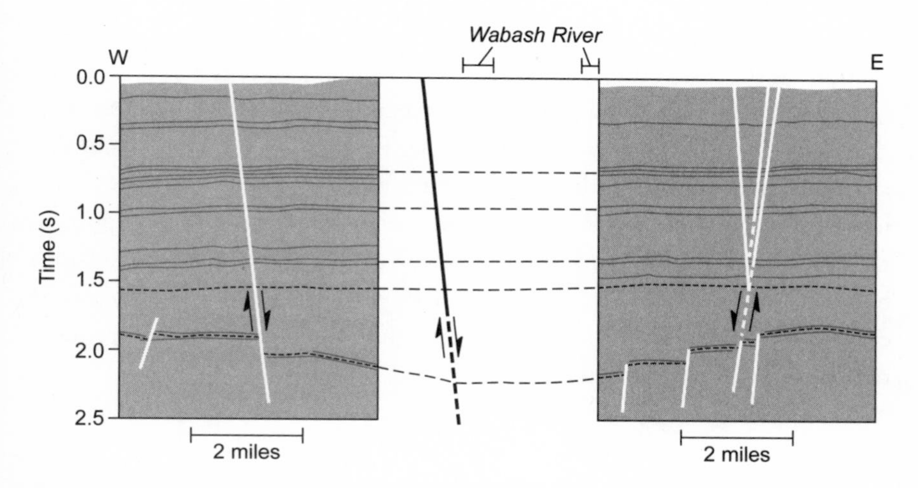

FIGURE 11.3 A seismic section across the Southern Indiana arm of the Reelfoot rift. (After Sexton et al., 1986)

of the area from rocks at the surface and from drilling shows where different rock layers are. Interpreting seismic sections is an art practiced by geologists who know a lot about an area and make educated guesses that are generally pretty good.

This section crosses the rift, with a gap in the middle where data couldn't be collected in a town and under the Wabash River. The reflections are connected across the gap, as shown by dashed lines.

The first feature to notice is the deep reflector. Starting at the right, it gets deeper until the gap, and then shallower again to the left. This reflector is thought to be the "basement," or the top of the rocks that were pulled apart to form the rift valley. Breaks in this reflector, marked by white lines, show where the part closer to the middle is dropped down. These are thought to show normal faults that formed during rifting and might be where earthquakes happen today.

The shallower reflectors are sedimentary rocks formed by sediments that filled the rift over time. There are also faults in these layers. White lines in the picture show that some shallow faults connect with the deeper faults. This indicates that some of the motion on the shallow faults occurred because of motion on the deep ones.

The vertical offset on the deep faults gets up to 1,500 feet, so these faults show up very well. However, the shallower faults don't show up as well because the offset on them is less than about 300 feet. That means that over the hundreds of millions of years since the sedimentary rocks were deposited, the faults haven't moved much. This fits with the idea that the old faults don't heal and so move occasionally when forces within the continent get strong enough to revive them. This seems to have happened several times, as mentioned in the previous chapter.

More detailed seismic studies of the sediments close to the surface also show faulting within the past 10,000 years. This is the time period that matters for thinking about today's earthquakes.

Beyond using seismic waves, more can be done to study what's underground. With the class "mystery box," additional information comes from weighing the box and checking it with a magnet. For the New Madrid area, measuring gravity and magnetism lets us "see" deeper into the earth than the seismic section. The seismic section stops at the top of the basement rocks because the waves sent down didn't have enough energy to return reflections from deeper depths.

Underground rocks are weighed by measuring the pull of gravity. Less dense rocks pull less, so you weigh less when you walk over them. You don't lose much weight—only about 1/100,000 of what you'd weigh if the rock were denser. Gravity meters can measure these incredibly small changes. That's useful because the sedimentary rocks filling the embayment are less dense than the rock in the surrounding crust.

Gravity above the rift is less than on either side due to the low-density sediments. Because the seismic section shows how thick the sediments are, it's possible to calculate how much less the gravity should be. Taking this into account shows an interesting result: Gravity across the whole embayment is higher than on either side. This means that under the sediment there's denser rock, which pulls harder. That's why in the model in figure 11.1 there's a "pillow" under the rift of lower crust that is denser than the upper crust. Researchers think that this pillow probably formed during the rifting.

In the same way, magnetic measurements at the surface tell about the rocks below. Igneous rocks like granite or basalt, which are formed by volcanic processes, contain more magnetic minerals than sedimentary rocks. Within the embayment, there are places where the magnetic field is stronger, so researchers think there are blobs of igneous rock down there that were intruded from below into the surrounding rocks. It's not known whether these formed during the rifting or later.

Real-world Messiness

Models like the one in figure 11.1 are useful as long as we remember that they're simplifications of the complicated real world. To decide how much to believe a model, it's important to think about what went into it and how good these assumptions were. There are some important issues to think about.

First, it's always possible to make different interpretations of data. Many different structures down in the earth look similar when geophysicists try to figure them out from data at the surface. For example, gravity shows that there are dense bodies at depth, but it's hard to get their shapes right. Larger dense bodies can look a lot like smaller and even denser ones. It's also hard to tell how deep something is because a shallow dense body can look a lot like a deeper and denser one. There's a way to get around this somewhat, because deeper bodies affect the gravity over a larger area. Another way is to combine different kinds of data. In particular, because nothing in the seismic section looks like a piece of dense material, the dense stuff is probably below the deepest depth in the seismic section. Thus, the rift model is a reasonable idea of what's down below, but the details aren't known and important pieces of the picture might be missing.

Second, the rift model comes from combining data from different places. The seismic section we looked at is from the Southern Indiana arm, not the Reelfoot rift itself. This makes the model in figure 11.1 an average over different places that aren't exactly the same. For example, there are differences between the Southern Indiana arm and the Reelfoot rift. This isn't surprising because the area north of the rift has a geologic history in which events other than the rifting have been important.

Third, the model relates the deep structure of the rift to the faults where earthquakes happen today, either on the old faults or younger, shallower ones. That seems reasonable in the Reelfoot rift, which is the most active part of the New Madrid seismic zone. The situation is more complicated with the three northward arms. These arms are useful as a general idea, but it's not clear how the faults that show up in the geology and the locations of the earthquakes relate to the deep structures. It looks like the concentration of earthquakes and faulting in the Reelfoot rift spreads out into a broader and more complicated region of smaller earthquakes and faulting. Some of these faults formed as part of the rifting, and others formed at different times and in different ways. This makes understanding what's going on there even harder than at New Madrid itself.

Why New Madrid?

Although the model gives some ideas about what's happening to cause the New Madrid earthquakes, it doesn't help us with two big questions.

The first question is why New Madrid seems to have more large earthquakes than other parts of the continental interior. Small earthquakes happen in lots of places—even near Chicago—but at least in the short time over which we have data New Madrid and its surroundings have had more large earthquakes. Figure 11.4 is a map compiled by Steve Marshak of the University of Illinois that shows that the interior of the continent is full of old failed rifts and old faults. We haven't seen large earthquakes on most of these. That

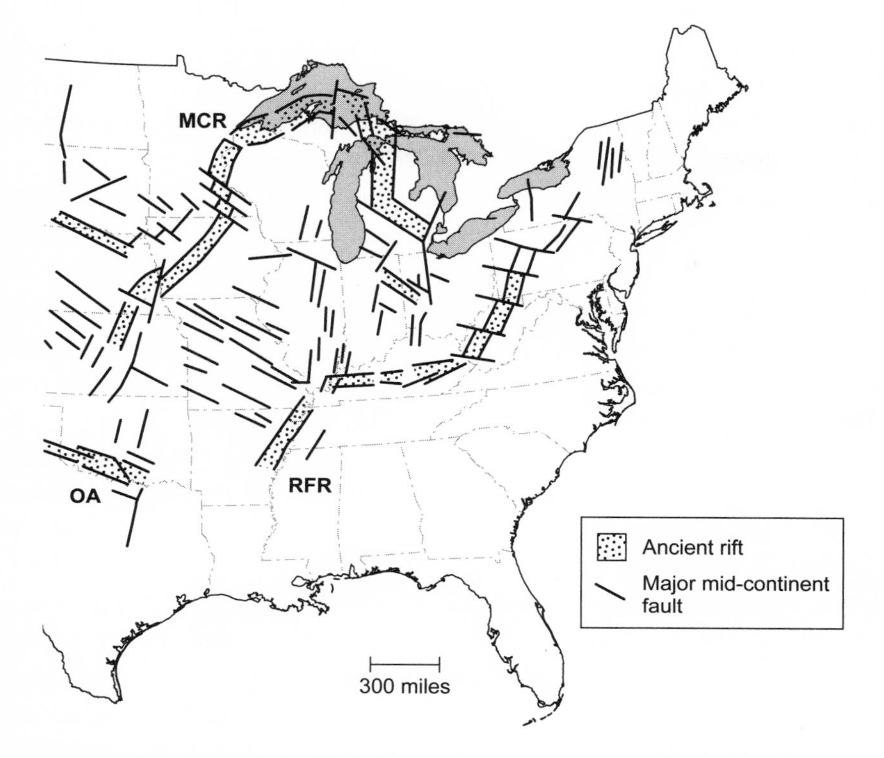

FIGURE 11.4 Map of faults and ancient rifts in the central and eastern United States. Labelled structures are the Reelfoot rift (RFR), Mid-continent rift (MCR), and Oklahoma aulocogen (OA). (After Marshak and Paulson, 1996)

leads to the question: If the Reelfoot rift is weak after all these years, these others could be as well, so why don't they have earthquakes? In particular, the Mid-continent rift that runs from Lake Superior to Kansas is older than the Reelfoot and is a much bigger deal geologically. It jumps out on U.S. gravity and magnetic maps while the Reelfoot barely appears. There's also the southern Oklahoma aulocogen, another failed rift that formed as part of the same rifting that caused the Reelfoot.

This means that saying the New Madrid earthquakes occur because the Reelfoot rift is there is true but only part of the story. It's like the joke where someone traveling in a hot air balloon is lost, asks someone on the ground "Where am I?" and is told "You're in a balloon."

The second question is what causes the earthquakes. The world is full of faults, but earthquakes only happen when something makes them move. Earthquakes happen on the San Andreas fault because the Pacific and North American plates on the opposite sides of the fault are moving in opposite directions. Earthquakes at New Madrid and other places within the North American plate show that some process or combination of processes is causing small motions inside the plate. These motions are a lot smaller than the motion of the whole plate but are big enough to cause earthquakes. The plate is like a big chunk of floating ice with small cracks in it; the whole chunk moves but there are also small motions across the cracks.

If something is going on inside the North American plate, what is it and why? At this point, we don't know. As is often the case in geology, there are a number of explanations, each of which has strengths and weaknesses, but none is strong enough or supported by enough data to explain all the observations. Perhaps there's no single cause, and several effects contribute to what's happening.

One possibility is that forces due to plate motions cause the earthquakes. The North American continent is part of the North American plate, which extends all the way to the Mid-Atlantic ridge. At the ridge, new rock is added to the plate. Over time, this slow addition has made the wide Atlantic Ocean. As this hot rock cools, it gets denser and so moves away from the ridge. This cooling causes a force within the plate called "ridge push," which is part of what makes the plate move. Computer models predict that this push should be transmitted into the continent.

There are also forces acting on the bottom of the plate that are caused by mantle flow. Alessandro Forte of the University of Quebec has done a computer model of the flow and found a surprising result. The small Juan de Fuca plate subducting under Oregon and Washington, shown in figure 9.7, is what's left of a much bigger plate that has been subducting for about 100 million years. Mantle material dragged down by the subducting plate is heading east under the central U.S. and meeting material going west from the Mid-Atlantic ridge. This collision builds up stress on the bottom of the plate under a wide region of the Midwest.

Another possible source of the forces that move the faults is the great mile-thick ice sheets that covered much of North America during the last ice age. These ice sheets shaped the landforms of the Midwest, including the Great Lakes. Although the ice sheets started melting about 18,000 years ago, the land is still moving as a result. Small motions, called "post-glacial rebound," happen because the mantle below the earth's crust flows like a very gooey fluid—much stickier than road tar or maple syrup. The weight of the ice sheets pushed material in the mantle beneath them away, and it's now flowing back. We'd thought that this was happening, so it was great when GPS came along and let a group of us map it. It was fun to see the vertical motion today caused by the long-gone glaciers (fig. 11.5). For the first time, we had data to test ideas about where and how fast this motion is over a huge area.

This result got a lot of press coverage in Canada because it shows the GPS sites in Canada rising and the ones in the U.S. sinking. However, Giovanni Sella, the lead author who did the work as a postdoc at Northwestern, had a bit of trouble getting permission to publish it after he joined the U.S. National Oceanic and Atmospheric Administration. Because this agency does Great Lakes studies, I joked that he should explain that post-glacial rebound is good for the U.S. We gain valuable water as the northern—Canadian—shores of the Great Lakes rise and the southern—U.S.—shores sink. In exchange, Canada gets better beaches. This impacts industries and homeowners along the shores of the Great Lakes as well as international management of water levels and shipping.

I've been interested in post-glacial rebound for years and have written papers arguing that it causes earthquakes, like the one in 1929 on the Grand Banks, along parts of the east coast that were covered by ice sheets. The idea is that the crust was bent under the weight of the ice and is experiencing "unbending" forces since the ice melted. These forces are causing motion on the old faults left over from when the continents rifted to open the Atlantic Ocean. The GPS data support this idea, because they show large motions along Canada's east coast.

A similar process is a likely cause of the Saint Lawrence River valley earthquakes, as shown by GPS data acquired by Stephane Mazzotti and coworkers from the Geological Survey of Canada. It helps that the Saint Lawrence is near the "hinge line," where rebound motions change from up to down.

However, the GPS data (fig. 11.5) don't show anything special happening at New Madrid. This observation agrees with computer models that predict that because this area is farther south, post-glacial effects should be small and dying off with time. If this process causes earthquakes, they should be more common north of New Madrid. The fact that we don't see this implies that post-glacial rebound alone isn't causing the New Madrid earthquakes.

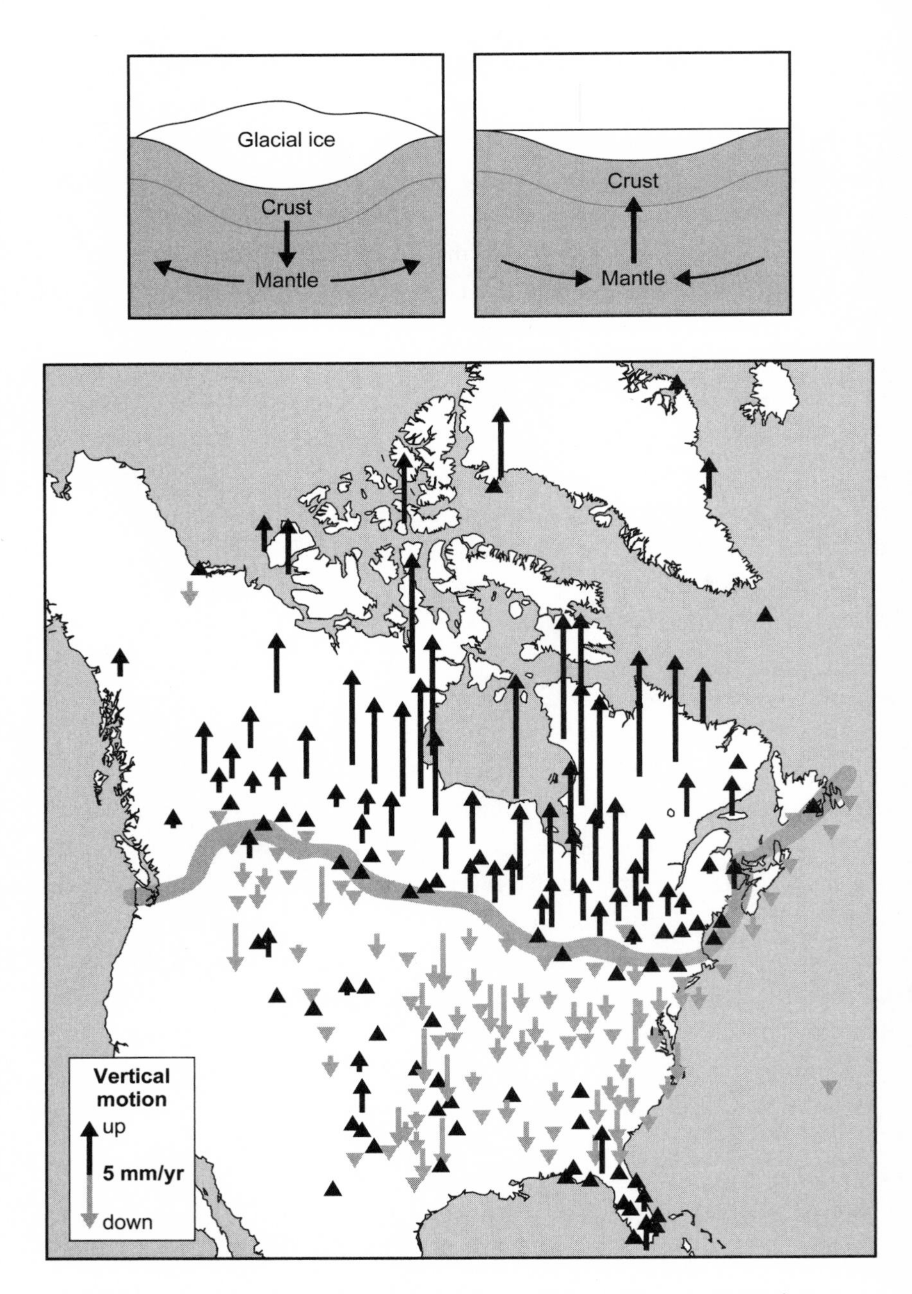

FIGURE 11.5 *Top,* How post-glacial rebound occurs. *Bottom,* GPS data showing vertical motion due to post-glacial rebound. (Sella et al., 2006)

Plate motion, mantle flow, and post-glacial rebound are large-scale processes that affect most of central and eastern North America. Thus although, they may contribute in a general way to the forces that cause the New Madrid earthquakes, they don't explain why in the past few hundred years large earthquakes seem to happen more at New Madrid than other places.

Some proposed explanations assume that there's something geologically special about New Madrid. Either forces acting there make it special, or the rocks there are different. Both of these ideas are possible because the Reelfoot rift is different from its immediate surroundings.

Lanbo Liu and Mark Zoback of Stanford have argued that the rocks under New Madrid are hotter than those in the rest of the central U.S. and so can slide more easily. However, that doesn't seem right. I've studied this with two experts on temperatures in the earth: my wife, Carol, a geologist at the University of Illinois at Chicago, and Jason McKenna, from the Army Corps of Engineers. From the few measurements available, we found that New Madrid doesn't look hotter than its surroundings and that the earlier result had several problems including a mistake in a computer program. To be fair, such things happen to all of us.

Another idea is that the high-density pillow of lower crust under the rift is sinking and making the faults move. The problem with this explanation is that the pillow has been there for hundreds of millions of years, but the faults haven't moved during most of this time. For this idea to work, something unknown must have happened recently to let the pillow sink. Those who like the idea suggest that the area became hot and weak recently, but our study didn't see any such heating.

A similar idea, by Paul Segall of Stanford and Shelley Kenner of the University of Kentucky, is that the earthquakes occur because some unknown weakening process happened at the right time to make it easier for the faults to slip. The difficulties with this explanation are that there's no reason to believe the weakening happened and it predicts motions that the GPS data don't show.

A stronger candidate, in my mind, is the idea that the forces are due to changes in the amount of sediment in the Mississippi River valley. Roy Van Arsdale of the University of Memphis has proposed that over the past 20,000 years, the erosion of a lot of sediment would make the crust "unbend," just the way it did when the glaciers melted. Andy Freed and Eric Calais of Purdue have made a computer model of this process and found that these forces could be big enough to cause earthquakes.

The challenge for all of these explanations is that earthquakes have happened both in other places in the North American continental interior at other times and in the interior of other continents. This means that either there's a general explanation for mid-continental earthquakes, or the earthquakes in each place result from special local conditions. Moreover, as we'll see, earthquakes at New Madrid and other places sometimes go on for a while and then stop for a while. As a result, focusing on the places where earthquakes happen to have occurred in the past few hundred years can mislead us and make us miss the big picture.

Chapter 12

Guidance from Heaven

Whenever a new discovery is reported to the scientific world, they say first, "It is probably not true." Thereafter, when the truth of the new proposition has been demonstrated beyond question, they say, "Yes, it may be true, but it is not important." Finally, when sufficient time has elapsed to fully evidence its importance, they say, "Yes, surely it is important, but it is no longer new."

—Montaigne, 1533–1592

The rift model for New Madrid developed in the 1980s was a huge advance. Still, seismologists couldn't say whether a large earthquake like those of 1811–1812 would happen in the next few hundred years. All we could say was that there had been earthquakes in the past, so they might happen again.

The GPS Revolution

Since the 1980s, an incredible, almost magical, new technology has changed everything. The magic wand is GPS, the Global Positioning System. GPS sounds too good to be true. Simple GPS receivers that are the size of a cell phone—or even inside a cell phone—give location to an accuracy of a few feet. The fancier ones used in geology locate a point on earth to a millimeter, or 1/25 of an inch.

A GPS receiver analyzes radio signals sent from satellites orbiting 12,600 miles above the earth. Using the fact that radio waves travel at the speed of light, the receiver figures how far away each satellite is and thus where the receiver has to be for each satellite to be that distance away. This is like locating earthquakes using signals from one source received at many seismometers, except that GPS uses many signal sources that are received at one place (fig. 12.1).

The way this works is neat. The radio signal from each satellite identifies the satellite and describes its orbit. The signal also gives the time when it left the satellite, as determined by an incredibly precise atomic clock on board. This time signal looks like a set of pulses. The GPS receiver also has a clock and a computer inside, so it generates its own set of time pulses and matches them to the ones coming from the satellite to find how long the

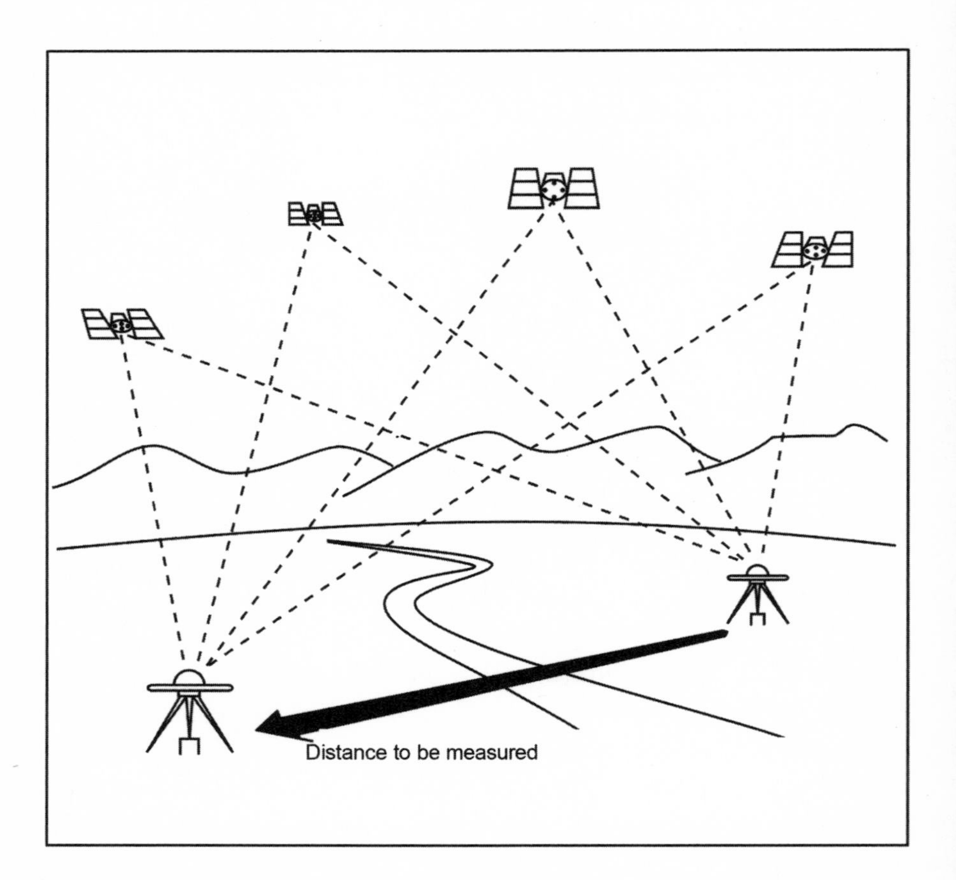

FIGURE 12.1 Using precise positions of GPS receivers determined by tracking GPS satellites, measurements over time show how sites are moving relative to each other. (Stein and Wysession, 2003)

signal took to arrive. Using this method gives the receiver's position to an accuracy of about 10 feet.

Getting more accurate positions is harder. It uses the fact that the satellite sends the time signal by adjusting, or modulating, a "carrier" radio signal. That's actually the way all radio signals are sent. Radio stations are designated as AM or FM. These stand for "amplitude modulation," where the height of the carrier wave is modulated, and "frequency modulation," where the period of the carrier wave is adjusted. Because the carrier wave has a shorter period than the time signal, matching the carrier wave from the satellite to one generated by the receiver gives a position accurate to fractions of an inch.

Although the GPS concept is simple, the details are incredibly complicated. Radio waves travel 186,000 miles in 1 second, so they need only one-ten-billionth of a second to travel an inch. That means the GPS satellite clocks have to be incredibly accurate. Radio waves are slowed by water vapor in the atmosphere, so a correction is needed. The orbits of the GPS satellites have to be known very accurately because any error in their positions can make it look as if the ground is moving. Even Einstein's theory of relativity has to be factored in because it shows that the clocks in the fast-moving satellites run more slowly than ones on earth. This is like the idea that a space traveler who travels to a far-away solar system comes back younger than friends left on earth.

Amazingly, all of this works. GPS is a perfect example of Arthur C. Clarke's famous principle that "any sufficiently advanced technology is indistinguishable from magic." Clarke, best known as a science fiction writer, knew what he was talking about. He wrote about using satellites to communicate between points on earth years before anyone put satellites in orbit.

GPS has changed the world in many ways. Throughout history, finding a position on earth was a huge challenge. Now, people push a button and know where they are more accurately than ever previously possible.

In fact, GPS works better than the people who designed it planned. It was developed by the U.S. Defense Department, whose scientists wanted it to be accurate to about 100 feet. They included special features in the satellites' signals that made them less accurate except when using special military equipment. However, during the 1990–1991 Persian Gulf War, U.S. soldiers bought small, lightweight GPS Receivers (SLGRs), or "sluggers," because they hadn't been given military receivers, so the government stopped trying to degrade the signals. This never mattered for science because geodesists had already figured out methods to get positions to fractions of an inch.

GPS and similar methods of finding positions on earth using observations from space are called space-based geodesy. They give us a new view of the world, which is one of the biggest changes in geology during my career. Geologists used to think in slow terms. We studied how plates moved by looking at seafloor magnetic stripes formed over millions of years and measured how faults moved by looking at features that had been shifted over thousands of years. Suddenly, we were doing the geology of now, seeing the world change before our eyes. We could watch the Atlantic Ocean grow an inch wider over a year.

I got involved with GPS after coauthors and I developed the NUVEL-1 model that showed how plates had moved over 3 million years. Models like NUVEL-1 gave scientists developing GPS and similar space-based methods a way to check their answers. This was important because otherwise they would have no way to tell if the speeds and directions of motion coming out of these complicated technologies were right.

The fact that GPS and plate motion models gave very similar answers had a major impact on geology. Beyond proving that GPS worked, the agreement showed that plate motions were steady over time; motions over a few years are about the same as over a few million years. Plates move steadily, although the faults at the boundaries between them are locked and only move occasionally in big earthquakes that release the plate motion that has accumulated since the last one. The occasional motion on the boundaries is damped by the fluid asthenosphere under the plates, leaving smooth motion between points within plates. This is like the way a car's shock absorbers dampen the effect of bumps on the road, giving a smooth ride.

GPS changed earthquake studies dramatically. Before it, seismologists used seismograms to study earthquakes when they happened, but we couldn't do much before they happened. We knew that rocks on either side of a fault should be moving slowly as strain that would be released in the upcoming earthquake was stored. The problem was that measuring this motion with traditional surveying methods was so labor intensive that it rarely was done. Suddenly, GPS lets us do it anywhere, easily, and cheaply. Geologists put markers in the ground, measure their positions over a few years, and observe the ground moving. The ground movement tells us that an earthquake is coming.

Starting around 1985, faults all over the world began sprouting GPS antennas. Within a few years, these studies showed strain building along the

San Andreas at the rate we expect from the geology there. A few years of GPS gave the same result as the measurements over thousands and millions of years that we talked about in Chapter 9.

Even better, GPS showed movement as strain built up in places where we don't have historic records of earthquakes, but think from the geology that there have been large earthquakes in the past. Geologists suspected that big subduction zone earthquakes might occur along the coasts of Oregon and Washington—and now we could see the strain accumulating for the next earthquake. GPS also showed strain accumulating across the Wasatch fault near Salt Lake City, in Nevada, and in many other places. Whether on or off plate boundaries, there was motion everywhere we expected earthquakes. To our surprise, New Madrid turned out to be very different.

Interestingly, the new technology didn't arise because one person suddenly made a great discovery. Hundreds of scientists and engineers were involved, and the technology sort of grew slowly. When I started going to NASA meetings, the idea was "pie in the sky," something that might some day work. Although the technology folks were confident, geologists were hopeful but skeptical as we sat through long meetings full of organizational charts and jargon. We knew how complicated the earth was and that many clever ideas that ought to work didn't. Still, over the years space geodesy went from ideas, to development, to tests and then finally there were answers that mattered. The technology folks had delivered an incredible new geologic tool.

Fun in the Field

GPS was so much fun that I wanted to try it. Although GPS projects were starting in geologically interesting areas around the world, there were still places open. One was the Andes Mountains, where GPS measurements could show how subduction beneath South America was building this huge mountain chain. Another was New Madrid, where GPS could show whether to expect a big earthquake soon. I'd wanted to get involved with New Madrid since I arrived at Northwestern but couldn't think of anything useful to do that someone else wasn't doing. This seemed like a good chance though I didn't expect much to come of it. I certainly never expected that New Madrid would become one of my main research areas, much less that students I advised or taught would become New Madrid researchers at other universities.

There were two problems. First, I didn't know how to make GPS measurements. Second, I didn't have any GPS receivers. My situation was typical of geologists interested in these kinds of problems; GPS was so new that most of us were in the same boat.

Fortunately, there was help out there. Tim Dixon from NASA's Jet Propulsion Laboratory (JPL) was also interested in these problems and had the GPS expertise. JPL was a leader in developing GPS technology, and Tim was one of the first geologists to apply the new technology. He'd already had major successes that showed how much could be done. The expensive, high-precision GPS receivers could be borrowed from the UNAVCO, a consortium of universities that bought these receivers with National Science Foundation and NASA funding and helped researchers use them.

UNAVCO, whose motto could be "to slowly go where no one has gone before," was making the new technology a powerful research tool. I went to UNAVCO's facility in Boulder, Colorado, and learned how to use the equipment from their helpful and professional staff. I was so impressed that in 1999–2000 I spent two years as UNAVCO's scientific director and led the effort that got them funding to install a huge GPS network across the entire western U.S.

Along with Tim and Joe Engeln, a former Ph.D. student of mine who was on the faculty at the University of Missouri and is now assistant director of the Missouri Department of Natural Resources, I put together a project plan and got NASA funding. Thus, while Iben Browning drew national attention to New Madrid, we were getting ready. This was my first venture into big-project science. It was fun but scary. I'd always worked with a small number of people using data and analysis methods that I could personally control or check. This was going to be very different. By the time the project finished in 1999, more than 70 people from nine institutions had been involved.

In spring 1991, we drove through parts of six states and visited the town of New Madrid, where we bought T-shirts and hats. Our goal was to find places like state parks or roadsides where we could put markers and measure their positions. Good markers were crucial for the study because we wanted to measure geological motions, not small shifts of the markers themselves. Building these markers was a new art, and people were experimenting with different designs. Mike Bevis from North Carolina State University taught us how to install markers in solid rock, and Ken Hudnut of the USGS showed us how to drive 10-foot steel rods into the ground at most of the sites, where there wasn't solid rock (fig. 12.2). The hardest worker in this effort was John Weber, a

FIGURE 12.2 John Weber, Richard Sedlock, Joe Engeln, and I installing a GPS marker at Malden, Missouri.

talented young field geologist who would use the study for his Ph.D. thesis. We carefully picked a mix of 24 sites to make sure that if anything were happening, we'd see it. Some were in the most active part of the seismic zone, where motions might be greatest, but the markers were in soft dirt and might show spurious motions. Others were in a ring around the seismic zone, farther away but in solid rock.

Researchers from many Midwest institutions were interested in trying to see if the new technology could help us understand New Madrid. We got together in November 1991 at the University of Memphis. This might seem funny, given the later disagreement between Memphis scientists and us about what the data we collected together meant, but it shows one of the strengths of science. We were going to measure what was happening and make all the data and results public. This meant that anyone involved—or any other scientists for that matter—could make their own interpretation of the data.

The fieldwork was hectic and lively, like a summer camp for adults. It helped that most of the group were Northwestern students who were helping out for fun and for a good excuse to get away from class. As well as learning to set up and operate the equipment, we played Hacky Sack and sampled Memphis's famous barbecue. The only downside was that we'd planned the survey for November, when there should have been little rain and comfortable temperatures. Instead, there was record cold for the area, and it was cool even by Chicago standards. Fortunately, an outdoor equipment store was having an end-of-season sale, so we bought a lot of cold-weather clothing at a good price. Otherwise our planning, which went as far as giving everyone bright-red hats because it was hunting season, worked well.

After a few days, everyone was ready (fig. 12.3). We moved to a motel in Hayti, Missouri, 22 miles from the town of New Madrid. It's at the junction of I-55, which runs north-south, and I-155, which crosses the Mississippi River into Tennessee. That let us work at many of the sites during the day and get together at night. For the next five days, everyone drove out before dawn, set up the GPS antennas at the markers, and then relaxed while the GPS receivers collected data for 10 hours. It was a comfortable routine. We'd rented minivans and removed the middle seats, so on cold days we kept the receivers in the vans. On nice days, we set them up outside, read, relaxed, wrote, or slept (fig. 12.4).

There were the usual amusing incidents of any field program. The best was when someone didn't keep track of his expensive GPS receiver, so a group "stole" it, hid it, and returned it only after the resulting panicky search. We

FIGURE 12.3 Some of the first survey gang. *Front*, Tony Thatcher, John Weber, Mark Woods, Joe Engeln, Brennan O'Neill, Lynn Marquez, Wai-Ying Chung, Ann Metzger; *Back*, Aristeo Pelayo, Dave Tralli, Steve Vasas, Richard Sedlock, Dave Bazard, Jonathan Rich, Harvey Henson, George Viele, Eric Kendrick, Bob Smalley, Lisa Leffler, Cristina Reta, Gary Acton, Seth Stein.

visited local landmarks, including a restaurant in nearby Sikeston, Missouri, that's famous for throwing rolls to patrons. Once we finished with these sites, we split up and moved to the more distant sites. Often, people stopped to ask what we were doing, leading to fun chats and more. Several people offered us snacks, and one student working at a site by an airport managed to cage a free airplane ride. We went back to Chicago with warm feelings and gratitude to the people of the area.

Our plan was to go back in two years, when the ground might have moved enough for us to see it. Competing researchers from Stanford—Paul Segall, Mark Zoback, and Lanbo Liu—took a different strategy. Rather than put in new markers and measure them with GPS, they used GPS to resurvey markers that had been surveyed in the 1950s with older methods. That was a high-risk, high-reward strategy. If it worked, they'd see ground motion before we did. The risk was that the results could have large errors. The old markers hadn't been designed for the kind of accuracies needed today. There was also a bigger problem. GPS data are analyzed so that bad data at one site are easy to spot and discard because the site seems to be moving in a different way

FIGURE 12.4 Steve Fisher from JPL hard at work during the first survey, showing why GPS can also stand for "Great Places to Sleep."

from the others. The old-markers approach mixes the data, so one bad site can make all the others seem to be moving. That seems to be what happened in an earlier study by one of the Stanford researchers who reported that southern New York and Connecticut were deforming as fast as the San Andreas fault, which was hard to believe and drew criticism.

In 1992, the Stanford group announced that they'd found "unusually high strain rates" in the southern part of the area. That's what's expected if a big earthquake were coming, but the rates were much higher than expected, about one-third of the rate along the San Andreas. They decided to push the results hard. Scientists always gamble between going with new results or waiting until they're certain. There's no right answer: move too fast, and you might be wrong; wait too long, and you might be scooped. Most scientists have erred both ways. We're happy about the times we gambled right and chagrined about the times we didn't. There's the added complication that we're naturally quickest to believe results that match our expectations.

Many scientists, including me, were skeptical of the Stanford results. The rates seemed much too fast. Moreover, a similar study by Richard Snay from the National Geodetic Survey and Jim Ni from New Mexico State University had found no motion in the northern part of the area. Still, the Stanford authors seemed confident. When one of them told a conference that their New Madrid result was "the most important result you'll hear at this meeting," the person sitting next to me whispered "assuming it's true."

All we could do was stay with our plan. In October 1993, it was time to survey the network again. Most of this group were University of Missouri students, and the weather was unseasonably warm. There were the usual minor problems. One site was now under a newly constructed baseball field. Another took me hours to find because we'd mapped its position relative to a state park's dirt road, and the road had been relocated.

All of this was so exciting that Joe Engeln and I nearly ran out of gas while driving along and discussing the science. Soon we would have positions for the sites spaced two years apart, so we could see if anything was moving. John Weber took the data to JPL and after a few weeks had the answer—maybe. The sites were moving either much slower than the Stanford group thought or not at all. Within the uncertainty of the measurements, we couldn't tell.

Nothing Is Moving

We went back in fall 1997 to settle the question. By now, it all felt familiar. The students were from Grand Valley State University in Michigan, where John had become a professor after graduating. We loaded the receivers into the minivans and headed off. A few weeks later, Northwestern graduate student Andy Newman started analyzing the data, and a few months later he had the answer. Nothing was moving. He checked his results carefully, and the answer stayed the same: zero.

To understand the results, it's important to realize that when scientists measure anything, there's an uncertainty of measurement. No matter how hard they try, nothing can be measured perfectly. Any set of measurements is spread out because of all the little problems that arise in measurements. These are called "errors of measurement," but they don't mean that anything was done wrong. These are just the uncertainties of measuring anything in the real world.

For example, figure 12.5 shows a histogram of 50 GPS measurements of the latitude of "the rock" on Northwestern's campus where student groups have displays. The histogram is highest at the most common value and spreads out on either side of it.

Data like this are described using the famous bell curve, named after its shape (fig. 12.5). The height of the curve shows the probability that we'll get any specific value. The most likely value is called the mean, and the spread about the mean is described by the standard deviation. There's a 68% chance that any measurement will be within one standard deviation—above or below—the mean, and a 95% chance that any measurement will be within two standard deviations. Thus, a measurement is described as "mean plus or minus two standard deviations" and written with the plus-or-minus symbol, ±. The two-standard-deviation value is the uncertainty of the measurement. Real data aren't exactly described by the bell curve, but it's usually a good approximation and gets better as the number of data increase. Because it works so well for many kinds of data, the bell curve is technically called the normal distribution.

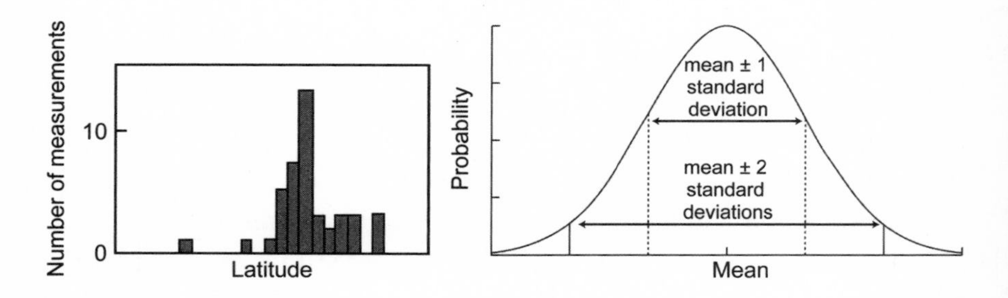

FIGURE 12.5 *Left,* Measurements of the latitude of a point on the Northwestern campus; *Right,* Most data are described by a bell curve using their mean value and standard deviation.

Understanding uncertainties is important for interpreting measurements. For example, think about the diet example from Chapter 1. To decide if your diet is working, you first need to know the uncertainty in the scale. You could estimate this by weighing yourself 10 times and using the spread of the measurements about the mean to approximate the two-standard-deviation range. If that uncertainty is 2 pounds, you describe your weight as mean value ± 2 pounds. Suppose one day it's 176 ± 2 pounds, and a day later it's 175 ± 2. Because the 1-pound difference is less than the uncertainty, there's no "statistically significant" weight loss. Perhaps the loss is real, but these measurements don't prove it. The way around this is to keep weighing yourself over a longer time. Even though the scale has the same uncertainty, you can tell after a couple of weeks whether you're losing weight.

GPS works the same way. Many complicated factors contribute to the uncertainty in computing a site position. These include the slowing of radio signals by water in the atmosphere, tiny motions of markers due to effects like changes in ground water, and uncertainties in the satellite orbits and clocks. Although a lot of effort goes into reducing the uncertainties, they're still not fully understood. Fortunately, we can reduce the uncertainties and get more precise measurements just by waiting. Because our goal is to find how fast a site is moving, a longer span of measurements gives a better handle on the speed.

Andy analyzed the data at the University of Miami, where Tim Dixon was now on the faculty and had established one of the world's best GPS labs. The extra years of data showed the rate of motion across the seismic zone as 0 ± 2 millimeters per year. Thus, the most likely rate of motion across was zero, and there was a 95% chance that the motion was less than 2 millimeters per year. Because a millimeter is 1/25 of an inch, that's very slow. Two millimeters is the thickness of a pencil lead.

Although our earlier data had been pointing to very slow motion, zero was a surprise. We had expected to see some motion. Instead, each time we measured, the motion came out slower. The time had come to accept that, to the accuracy of our measurements, the motion was zero.

This result showed that ideas about what was going on at New Madrid needed changing. Paleoseismic studies had found sand blows that scientists interpreted as evidence that New Madrid earthquakes comparable to those of 1811–1812 occurred in about 1450 and 900 A.D., or about every 500 years. Thus, a common idea was that the 1811 and 1812 earthquakes were great—

magnitude 8—and ones like them should happen again in a few hundred years. However, the GPS data showed that the ground was moving, or deforming, much too slowly for this to be true. At this rate, the most deformation the ground could store up in the next few hundred years would allow for a magnitude 7, not an 8. As we saw in Chapter 8, magnitude 7 is big but a lot smaller than 8. Thus we thought that the 1811 and 1812 earthquakes were probably magnitude 7, which is what Sue Hough soon showed from the historical accounts.

The best interpretation of the GPS data was no motion at all; the ground isn't storing up any deformation for a future earthquake. I thought the simplest explanation was that the recent sequence of earthquakes was ending. People had been thinking that the recent New Madrid earthquakes hadn't been going on for more than a few thousand years because the topography we'd expect from such motion doesn't exist. If the fault had turned on recently, why shouldn't it now be turning off?

Either way, it was clear that the argument the USGS was making—that New Madrid was more dangerous for earthquakes than California—wasn't right. To me, this wasn't a big deal. I'd never taken that argument seriously, and it wasn't till later that I learned that it had huge economic implications.

In December 1998, Andy presented our results at the American Geophysical Union conference in San Francisco. This annual conference is a weeklong geophysical circus, with tens of thousands of earth scientists presenting papers on hundreds of topics. Often, new results draw much discussion, so it was the perfect venue. The timing was also good. I was then UNAVCO's scientific director and keen to show other geologists how powerful GPS was.

Even by the standards of the exciting new GPS methods, the New Madrid data were surprising. Most GPS results from around the world confirmed what geologists already knew or suspected, but here, GPS was showing us something totally unexpected. GPS studies from around the world showed how the ground was moving, but here nothing was moving. We joked that these were the most boring GPS data the audience would ever see, but the story was in the boringness.

Andy prepared carefully for his talk because he was a graduate student expecting tough questions from listeners, including the Stanford researchers who were senior scientists. He didn't get much flak. Segall agreed that their earlier ominous finding "doesn't seem to be supported by the more recent results." A nice feature of science is that it's easy to back down when new and better data show you're wrong. No matter how good a job you do, someone will always come along and do better. We like being the ones who do better and accept it when in turn someone else does better than us. That's how science advances.

Our results got a lot of press interest because they contradicted the widely quoted claim that a big New Madrid earthquake was on the way. Academic science is a quiet, low profile affair that rarely makes the news. When our work draws interest, it's an opportunity to show the public what we do. Our results were slated for the front page of the *New York Times* science section, but got pushed to page five by the legalization of Viagra in Japan. Andy enjoyed having his research in newspapers nationwide but was disappointed that *Rolling Stone* didn't cover seismology. Still, he had a great thesis and is now a professor at Georgia Tech.

Naturally, USGS and CERI scientists who believed in a high earthquake hazard weren't happy. One told the news media, "It's irresponsible to make pronouncements about seismic hazard." Neutral observers, including ones within the USGS, cheerfully pointed out that the USGS had been doing exactly that for years. The critics also argued that we should "exercise great caution" in presenting results that differed from their claim of high hazard. I thought that they should have exercised caution before telling people in the Midwest that they faced a risk like California's.

Our results were soon confirmed. The Stanford group remeasured their sites and in 2004 announced that "in contrast to the 1992 study, we find no evidence for high strain rates." Another check came from Bob Smalley from the University of Memphis, who installed permanent GPS sites about where we had done our survey. Permanent, or continuous, GPS is much more expensive but better than what we did. The GPS antennas are mounted in the ground, which avoids the errors involved with setting them up each time. They acquire data all the time, not just every few years, so they measure the site motions more precisely.

In 2005, the Memphis group announced their results, which were pretty much the same as ours. Their motions were also tiny, less than 2 millimeters per year. The only difference was that they thought the motion between two of their sites was slightly greater than zero. This got press interest, especially because it was what FEMA wanted to hear. Three months before being forced to resign because of his performance during Hurricane Katrina, Michael Brown warned that "the entire Midwest is at risk."

However, separate studies by Eric Calais of Purdue, Charles DeMets from the University of Wisconsin, and Glenn Mattiolli from the University of Arkansas also examined the data and concluded that they didn't show any statistically significant motion. The small motion between two sites wasn't big enough to be convincing. Eric, who is very good at digging deep into data, even identified the small glitch that produced the apparent motion. Although it was hard to tell what caused the glitch, it's the size expected from the many possible sources of errors in the complicated GPS method. Problems like this often happen because important results are always at the edge of what the data might or might not show. The problems don't reflect too badly on Bob and his coworkers, who were providing and continue to provide high-quality data to the research community. Still, I suspect they now wish they'd checked this questionable result further before going public. As Eric told the *St. Louis Post-Dispatch*, "It's not fair in a scientific paper to scare people with things like that."

As more GPS data came in spanning longer periods of time, the uncertainty in the rate got smaller. As a result, the fastest motion possible from the data kept getting closer to zero (fig. 12.6). This happened whether the sites were in the New Madrid area or spread over a wider area of the central and eastern U.S. The data from successive studies show a clear trend that's exactly what we'd expected and impossible to dispute.

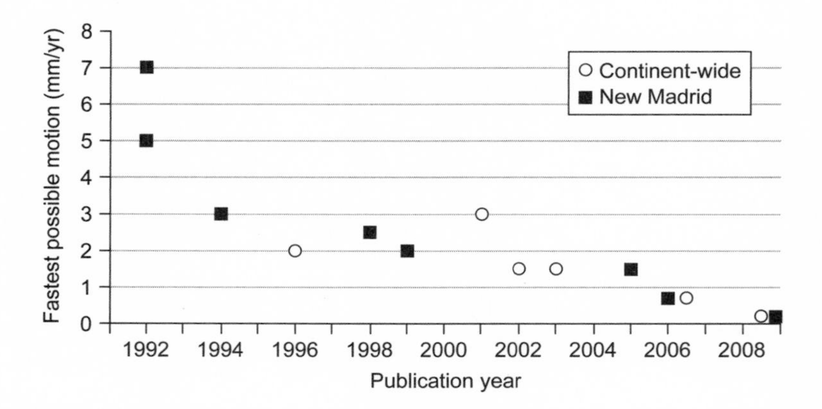

FIGURE 12.6 As the GPS measurements in the New Madrid area span longer times, the maximum possible motion they allow gets closer and closer to zero. (Calais and Stein, 2009)

In 2009, Eric's analysis of the most recent data showed that the motion is less than 0.2 millimeters per year (1/250 of an inch per year). That's the thickness of a piece of fishing line and 10 times smaller than the fastest motion allowed by the data in the 1990–1997 study. At this rate, it would take at least 10,000 years to accumulate enough strain for a magnitude 7 earthquake.

We're now sure that the ground in the New Madrid earthquake zone is deforming either very, very, slowly or not at all. A lot of thinking is going into what this means. Tim Dixon describes this debate by modifying Winston Churchill's tribute to the Royal Air Force in World War II, "Never in the history of geodesy have so many written so much about so little motion." Alternatively, in Shakespeare's words, there's much ado about nothing.

Chapter 13

Faults Turning On and Off

How wonderful that we have met with a paradox. Now we have some hope of making progress.

—Niels Bohr (1885–1962), a founder of quantum physics

The GPS data showing that the ground in the New Madrid seismic zone wasn't moving meant that we had to rethink ideas about Midwest earthquakes. We had hit one of the most exciting situations in science: a paradox where two different kinds of data seem to disagree. This means that at least one kind of data is wrong, or that both are right but need to be looked at differently.

One kind of data are the GPS results. Because year after year the data keep showing no motion, with even smaller uncertainties, it's hard to see how they could be wrong. Whether the data eventually show that the motion is zero or just tiny, the motion is much slower than it should be if large earthquakes are going to keep happening about every 500 years.

Could the other data, the paleoseismic observations, be wrong? Paleoseismologists have found sand blows and other liquefaction features that they interpret as showing that earthquakes as big as those in 1811 and 1812 occurred in about 1450 and 900 A.D. Field geologists are expert observers, so when they think they've seen something, they usually have. However, there can be other ways to interpret what they've seen. Because the dates of the sand blows have uncertainties, it's possible that instead of being produced by a few large earthquakes hundreds of years apart, the sand blows come from more frequent, smaller earthquakes. Another possibility is that large earthquakes around 1450 and 900 happened on other faults close to the ones that caused the 1811 and 1812 earthquakes, like those on either side of the Reelfoot rift. This would give paleoliquefaction features in about the same places, but the time between earthquakes on each fault would be much longer. Either possibility could explain why the GPS data don't show motion today on the faults that slipped in 1811 and 1812. However, paleoseismologists think these alternative explanations are unlikely.

The third possibility is that current interpretations of both the GPS and paleoseismic data are correct. That's what we decided to assume in 1999, when we presented the GPS data.

Although science is about making new discoveries, it follows some old traditions. Even in the Internet age, results aren't "official" until they appear in a scientific journal. Scientists try to put their best results in the most prestigious journals, like *Science* and *Nature.* Hence in April 1999 we published the GPS results in a *Science* article that distilled years of work down to the key result: The ground wasn't moving. We then explained what we thought it meant: If the ground continued not to move, the fault might be shutting down for a long time, so the earthquake hazard was a lot less than the government had been arguing. The time had come to rethink ideas of what was going on at New Madrid.

Changing the Model

As discussed in Chapter 11, scientists know that models aren't "right" but use them when they explain enough to be useful. The big question is what to make of the "anomalies," the data that a model doesn't explain. Because a current model is useful and familiar, we're reluctant to give it up.

Scientists' first instinct is to convince themselves that the anomalies aren't real or at least not so convincing that they need to be worried about. For example, current models of the brain don't explain telepathy, but researchers usually figure that's OK because the reports of telepathy aren't that convincing.

Sometimes despite all efforts to dismiss them, the anomalies still seem real. Our next instinct is to tweak the model by making small changes that explain the anomalies. Sometimes this works, but sometimes the patches get far-fetched. Recall in Chapter 9 how paleontologists found similar animals in different places and geologists came up with the idea of land bridges so they didn't have to give up the model that continents stayed fixed.

In these situations it's not clear which way to go. Both sides rely on Ockham's razor, the idea that the simplest model is best. Proponents of a model say that it's the simplest explanation and needs only what they see as minor patches. Opponents say the patched model is too complicated—"kludgy"—and a new alternative would be simpler. The argument can go on for years.

A bad sign is when the number of anomalies keeps getting bigger, so the model needs more and more patches. Eventually, some anomaly is so big that the model can't be usefully patched, and it's time for a new model.

That's what the GPS data did for New Madrid. Because the data show that the ground is moving so slowly that it would take tens of thousands of years to build up enough motion for a big earthquake, there's no easy way to keep the old model in which these earthquakes would keep happening every 500 years.

That simple observation was the straw that broke the camel's back. All of us interested in New Madrid knew there were many problems with the old model, but had been ignoring them. The GPS data made that impossible, so I thought it was time to discard the old model. This was easy for me because I had gotten involved in New Madrid studies recently and didn't have anything vested in the old model.

We needed a new model that would fit three puzzle pieces—GPS, seismology, and geology—together. The seismology and geology pieces were only partly new because some of each piece had been previously recognized. But, as so often happens, the whole was more than the sum of the parts.

The GPS results themselves were clear. The earth isn't storing up energy fast enough for large earthquakes to keep happening the way they have for the past 2,000 years. Although this surprised us at first, we quickly realized there wasn't any reason to expect big earthquakes to continue.

When I explain this to members of the public, they sensibly ask, "What about the little earthquakes we feel sometimes?" That's a good question with a surprising answer. In fact, although the small earthquakes generate media interest, scientists don't get too excited. That's because many of these small earthquakes look like aftershocks of the 1811 and 1812 earthquakes rather than signs that another big earthquake is coming.

Aftershocks happen after a large earthquake because the movement on the fault changed the forces in the earth, called stress, that act on the fault itself and nearby. Because parts of the fault slipped more than others in the large earthquake, there's a complicated, new pattern of stress on the fault. The stress has also changed on nearby faults. The jostling that results are aftershocks. Using seismograms from a big earthquake or by studying the fault at the surface, researchers can map the part of the fault that slipped and see that most of the aftershocks happened on it or nearby.

As discussed in Chapter 8, researchers have used the smaller earthquakes that have occurred since 1811 and 1812 to map the faults that they think broke then. This makes sense because the small earthquakes show patterns seen in aftershocks of large earthquakes in other places. Since 1812, the earthquakes have been getting generally smaller (fig. 13.1). By now, even magnitude 5 earthquakes are rare. This "decay," as it's called, happens because the stress changes produced by a big earthquake decrease with time.

The locations of the aftershocks also have a familiar pattern. Although none is very big, the largest ones (magnitude 5) are at the ends of what are thought to be the parts of the faults that slipped in 1811 and 1812. This pattern often appears on other faults because the stress left over from a big earthquake concentrates at the end of the fault. Thus although not all the earthquakes in the area are aftershocks—some occur fairly far from the faults that broke in 1811 and 1812—many seem to be.

It might sound surprising that aftershocks continue 200 years after a big earthquake. Actually, it isn't. In many places, aftershocks continue today from earthquakes long ago. For example, there are small earthquakes in Canada's Saint Lawrence River valley where a large earthquake happened in 1663. Mian Liu of the University of Missouri and I think there's a good reason why aftershocks go on longer within plates than on plate boundaries. On a plate

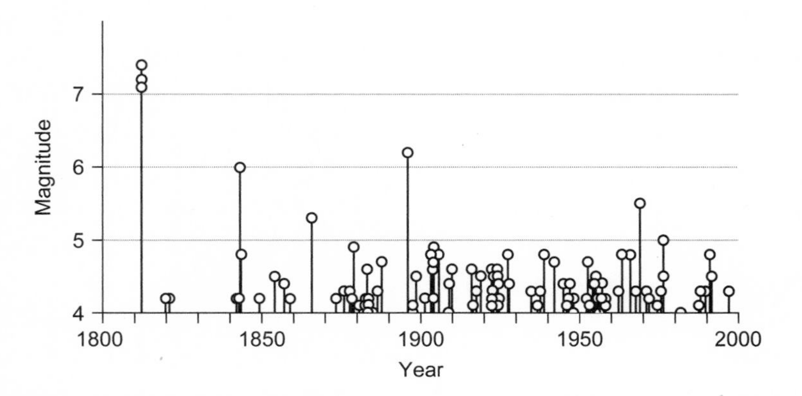

FIGURE 13.1 Earthquake history of the New Madrid area showing that since 1811–1812, the earthquakes have been getting generally smaller, as expected for aftershocks. (Stein and Newman, 2004)

boundary, the motion between the plates accumulates new strain on the fault quickly on a geological time scale. This "reloading" adds new stress that swamps the changes caused by the last big earthquake, so aftershocks are suppressed after about 10 years. Things are very different within plates. GPS data show that the interiors of plates deform very slowly, so after a big earthquake a fault reloads much more slowly than on a plate boundary. Thus, for a long time after a big earthquake, most of what happens on and near the fault results from the stress changes caused by the earthquake, so aftershocks continue for hundreds of years. It's like the way a hard-working crew cleans up quickly after a traffic accident but slower workers take all day.

To test this idea Mian and I used results from lab studies of faults to predict how much longer aftershocks would continue on slower moving faults. We then looked at data from faults around the world and found the expected pattern. For example, aftershocks continue today from the magnitude 7.3 Hebgen Lake earthquake that shook Montana, Idaho, and Wyoming 50 years ago. This makes sense because the Hebgen Lake fault moves faster than the New Madrid faults but slower than the San Andreas. The observations and theory came together the way scientists love but often don't get.

The third puzzle piece is the geology. Even before the GPS data, researchers knew that the New Madrid faults couldn't have been active the way they have been recently for very long geologically—more than a few thousand years. The motions in the 1811 and 1812 earthquakes can't be typical of what has happened on these faults for most of their long lives.

The simplest and most convincing proof is that there's none of the spectacular topography that would be there if these faults had moved a lot during their lives. The faults are hard to find on the surface. Even flying over the area at low altitude in a light plane, you don't see any sign of them. Because geologists use topography as a way of telling if the ground is moving, its absence was a major reason most geologists I talked to thought the GPS results made sense. As someone said to me, "If there were motions like on the San Andreas, there'd be mountains."

Another important clue is the shape of the faults. As figure 8.10 showed, the Reelfoot fault is a short "jag" that separates the Cottonwood Grove and North New Madrid faults. The jag is probably why there were three big earthquakes in 1811–1812 rather than one larger one. Over time, faults straighten themselves by breaking through jags like this. If the New Madrid faults had moved for millions of years the way they have in the past 2,000, the jag would long since have vanished.

The same idea comes from reflection seismology. As we saw in Chapter 11, reflection data show that although the faults are hundreds of millions of years old, most of that time they haven't been moving.

In summary, there were many reasons to think that the motion on the New Madrid faults had "turned on" recently. If faults turn on, they also have to turn off. The GPS data make it look like that's happening today.

Episodic, Clustered, and Migrating Earthquakes

The new results showed that we needed to think about New Madrid very differently. As Yoda explained in *Star Wars: The Empire Strikes Back*, we must unlearn what we have learned. In hindsight, this makes sense. Although New Madrid is in the middle of the North American plate, we'd been thinking about it pretty much like the San Andreas fault.

That was natural because many ideas about faults and earthquakes come from the San Andreas or other plate boundaries. The San Andreas is an earthquake-generating machine that has been running for millions of years and looks like it's going to keep going the same way. A lot is known about it. We know why it works: The North American and Pacific plates are sliding by each other. Although there are many faults in the boundary zone between the Pacific and North American plates, most of the action is on the San Andreas. We know that this has been going on for millions of years, because rocks on the two sides that used to be next to each other are now hundreds of miles apart. From the geologic record they left, we know that big earthquakes have happened every few hundred years for thousands of years. GPS shows that the plates are moving at about the same speed as they have for millions of years. Most important, GPS shows the earth storing strain across the fault. Because all the different kinds of data fit together, seismologists expect that in the next hundred years big earthquakes are very likely to happen on the segments of the San Andreas that broke in Southern California in 1857 and near San Francisco in 1906.

Without fully realizing it, seismologists had been acting Californian, applying California ideas to New Madrid. Although there are many faults in the central U.S., we'd been treating New Madrid as the only big player. We'd thought that because there had been large earthquakes there in the past, they would keep happening in the same place. Although we don't know why and how the New Madrid earthquake machine works, we figured it would keep working the same way for a long time. There wasn't any reason to think this, but before GPS we couldn't do any better. This assumption led to the idea that there was a large hazard from the next big earthquake on the faults that broke in 1811 and 1812 and a much smaller hazard from other faults in the region.

The GPS data showed that this model, summarized in the top of figure 13.2, was wrong. The new model is that earthquakes within the continents move around among many faults that turn on and off. These earthquakes are "episodic, migrating, and clustered." Earthquakes occur for a while on a fault system, which then shuts down for some time while other fault systems become active. This is like the way many paleontologists think evolution often occurs by "punctuated equilibrium"—long equilibrium periods during which species don't change much, punctuated by periods of rapid change.

Many geologists were skeptical when we proposed this idea in our 1999 paper, but the idea looks better all the time. Researchers are finding more and more data showing that faults in the New Madrid area turn on and turn off. John Holbrook of the University of Texas, Steve Marshak of the University of Illinois, and coworkers have interpreted the history of Mississippi River channels as showing evidence for earlier clusters of New Madrid earthquakes like the one that's been going on for the past 2,000 years. It looks like the clusters occur a few thousand years apart and last for a few thousand years. Researchers including Randy Cox and Roy Van Arsdale of the University of Memphis and Tish Tuttle of the USGS think that over the past 10,000 years, large earthquakes happened on other faults in and near the Reelfoot rift that are different from the ones that caused the 1811–1812 earthquakes. These faults don't have many earthquakes today but seem to have been active in the past. There's a complicated history that's just starting to be worked out.

There's also increasing data showing that faults in the mid-continent farther away from New Madrid have been much more active than they are today. Some seem to have had earthquakes that could have been as large as the New Madrid ones in 1811 and 1812. Although today only small earthquakes occur in southern Indiana, Stephen Obermeier of the USGS has found sand blows that seem to have formed in larger earthquakes, including a major one about 6,000 years ago. Another example is the Meers fault in Oklahoma,

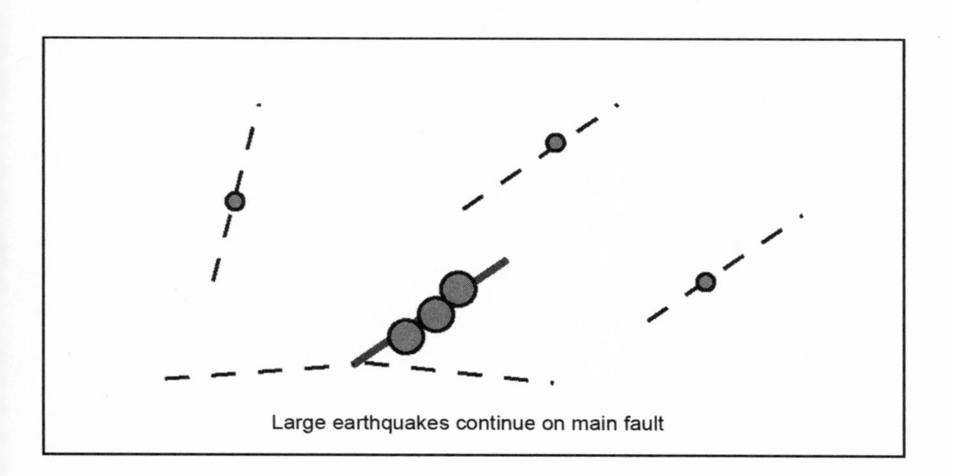

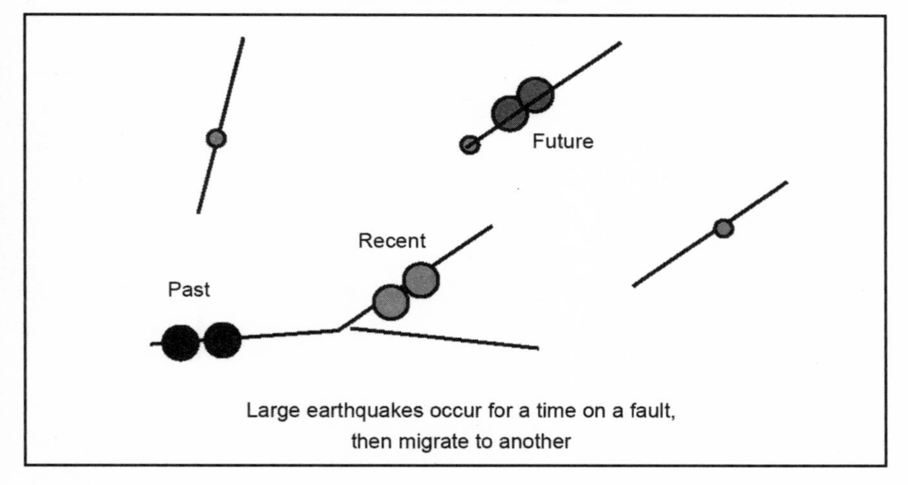

FIGURE 13.2 Ideas about New Madrid and other continental fault systems are changing from one in which one main fault stays active for a long time (*top*) to one in which many faults are active at different times (*bottom*). (McKenna et al., 2007)

which shows up as an impressive topographic ridge about 15 miles long and 15 feet high. Although there's essentially no seismic activity today, geologic studies found that it had large earthquakes within the past 2,000 years.

Similar patterns show up in other continents. Dan Clark and coworkers from Geoscience Australia have good data in Australia, where old fault features are preserved very well because there's not much rain. Even in the wet climate of northwestern Europe, Thierry Camelbeeck of the Royal Observatory of Belgium has found geologic evidence that earthquakes move around.

The best example might be in North China, which has a long earthquake record. Mian Liu and a Chinese colleague have noticed that since 1300 A.D., large earthquakes have been migrating so that no two magnitude 7 ones have happened in the same place. In fact, during the past 200 years a major fault system seems to be shutting down.

These other earthquakes show that although the forces acting within the different continents are different, the fault systems behave similarly. That makes sense. Even before I got involved with New Madrid, I'd studied enough earthquakes within plates to be skeptical of the idea that New Madrid was special. Geologists have learned over the years that few situations are unique. That's very helpful because most geological processes happen so slowly by human time scales that researchers see only a "snapshot" in any one place. To get around this, we combine information from similar situations at different stages.

The new model shows that earthquakes in the mid-continent can't be understood by focusing on one fault. Instead, geologists need to think about an entire system of faults that interact. Ross Stein of the USGS and his coworkers have shown that a big earthquake can change the stress on nearby faults and make it easier or harder for them to move in a big earthquake. This is called "unclamping" or "clamping" a fault.

To see this, think back to the soap bar and yoga mat in figure 6.5. Increasing the tension in the rubber band or reducing the weight of the soap bar unclamps the soap and makes slip easier. Conversely, reducing tension in the rubber band or putting weight on the soap clamps it, making slip harder.

Unclamping or clamping a fault can make a big earthquake happen sooner or later than it would have otherwise. As a result, stress transfer can cause earthquakes to migrate along a plate boundary. The best example is in Turkey, where since 1939 large earthquakes have occurred successively farther to the west along the North Anatolian fault. That's why in our paper after the great December 2004 Sumatra earthquake Emile Okal and I suggested—correctly—that a large earthquake would occur soon on the part of the plate boundary immediately to the south.

Fault interactions are even more important within a continental plate for two reasons. First, instead of a major plate boundary fault, there are many similar faults spread over a large area. Second, because faults within a continent are reloaded much more slowly than ones on a boundary, stress transfer after a large earthquake is more important.

Because of these interactions, faults within continents form what is called a complex system. Complex systems are ones in which the whole system behaves in a way more complicated than can be understood by studying its component parts. For example, the human body is more complicated than can be understood by studying individual cells, the economy is more complicated than can be explained by studying individual business transactions, and studying one ant doesn't tell us how an ant colony behaves.

Studying complex systems means doing science in a new way. Science traditionally works by focusing on a system's simplest components, understanding them in detail, and using the results to describe the whole system's behavior. In a complex system, that's not good enough. Instead, the system has to be studied as a totality, such that local effects in space and time result from the system as a whole.

Thinking of mid-continent faults as a complex system helps explain why we had made so little progress understanding earthquakes on them. Despite lots of looking, we hadn't found any real difference between the actual faulting process in earthquakes within continents and ones at plate boundaries. Why then did the earthquakes within continents have much more complicated patterns in space and time than those on plate boundaries? The problem was that we'd thought about individual faults, rather than thinking of them as interacting parts of a larger system.

The complex system idea also explains why there's such a complicated history of faults turning on and off. It doesn't look like there's a simple pattern in space and time. This makes sense because complex systems often give rise to chaos. As we talked about in Chapter 4 while discussing why earthquake prediction has failed, chaos is a situation in which changing a system a little will make it evolve very differently. As a result, predictions for the short term can be pretty good, but those for the long term aren't. That's why weather forecasts get worse the further into the future they go.

A nice analogy, suggested by Bob Smalley, is the game Booby-Trap (fig. 13.3). It's a box with a spring-loaded bar pushing on many pieces. When you remove one piece, some of the others slide around in a complicated way depending on the pattern of the pieces and how much each is being pushed. Some can be removed without making anything happen, but removing others makes many pieces slide. Removing a small piece can have a big effect, and removing a large piece can do nothing. All this makes it hard to tell what will happen before you remove a piece. The game is somewhat like earthquakes within continents, except that we don't have to watch for thousands of years to see what's going on.

Mian Liu, Qingsong Li of the Universities Space Research Association, and I have started making computer models of such fault systems. The models use a method called finite elements that was developed to help design things like buildings, cars, or airplanes. It represents the object as a grid with specific physical properties and calculates how it would respond when forces

FIGURE 13.3 The game Booby-Trap, an analogy for a complex system of faults.

are applied. Finite element models are used to study many geologic processes, including how faults behave. They let researchers look at systems too big for physical experiments (try getting the New Madrid fault system onto a lab table).

Although the mid-continent model is a huge simplification of a real fault system, it shows that earthquakes can cluster and migrate between faults. That's encouraging and makes it worth doing more detailed models. We and other researchers including Andy Freed and Eric Calais of Purdue are doing numerical experiments to try to understand how faults would work under different conditions. This will depend on the forces acting on the faults, the geometry of the faults, how easily they slide, and the earthquake history. The idea is to reproduce what the geology, earthquakes, and GPS show has happened and is happening. If that works, the models could give some ideas about what the future holds. Still, they're unlikely to predict exactly where and when the next large earthquake will happen, because that hasn't worked yet even in California where the fault system is much simpler and the forces are better understood.

One big question that the models let us explore is what forces within a continent cause earthquakes. Instead of being loaded at a steady rate by plate motions, intraplate faults are loaded at rates that vary with time by forces we don't yet fully understand. This variation is part of the reason the faulting has such a complicated pattern in space and time. In particular, the GPS data don't show the ground deforming anywhere in the mid-continent. It's important to figure out what this means about the forces.

Another question is why big earthquakes sometimes happen in clusters, like the New Madrid ones in 1811–1812, 1450, and 900. What makes clusters start and end? The numerical models show that once a fault has broken and relieved some of the stress that built up on it over the years, the next big earthquake should happen not on that fault but on other nearby faults that have had stress transferred to them. This result suggests that the simplest explanation for clusters is that the earthquakes before 1811–1812 were on different but nearby faults. If the paleoseismologists are right and the earlier ones were on the same faults, that could mean that the faults got weaker with time, making it easier for earthquakes to keep happening on them. Differences between faults might explain why some faults have clusters and how long clusters last.

An important aspect of the cluster issue is whether the fact that the GPS data show little or no motion means that the recent New Madrid cluster is ending. As Sarah Day of the Geological Society of London put it, it's the cluster's last stand. That's the simplest explanation, but there's an alternative. Perhaps some strain is left over from a process in the past, like the removal of the Mississippi River sediments. This strain would be dying off, but there might be enough left for another big earthquake. The question is like whether you can buy a Porsche every few years on a graduate student's salary. In the long run, the answer's no, but if you inherited some money, you could keep it up for a while. The longer the GPS data keep showing less and less motion the less likely another big earthquake seems, but the possibility still needs exploring.

Another part of this question is how stress is transferred between faults. If stress is migrating from the faults that broke in 1811–1812 to other faults, where and when should GPS data show strain accumulating? Should it be possible to see strain accumulation today? What should we look for and where?

Implications for New Madrid

The new data and model are a big step toward figuring out what kind of earthquakes to expect in the New Madrid zone.

Realizing that earthquakes within continents migrate and have long aftershock sequences explains why earthquake hazard mapping hasn't worked well in continents. Hazard mappers try to predict where large earthquakes will happen by looking at where small ones have. The locations of small earthquakes are pretty good predictors of where future small ones will happen. However, because large earthquakes sometimes migrate within continents, many small earthquakes show where large earthquakes have happened in the past, rather than where they will happen in the future.

A striking example is the way many scientists were surprised by the disastrous May 2008 magnitude 7.9 earthquake in Sichuan Province, China. The area had been mapped as having low hazard because there hadn't been many earthquakes on that fault in the past few hundred years. However, because earthquakes migrate, predicting big earthquakes based on small ones is like the carnival game Whack-A-Mole. You usually won't hit the mole by waiting for it to come up where it went down, because it will pop up somewhere else. However, geodesy should be much better for figuring out what's happening.

GPS measurements before the Sichuan earthquake showed about 2 millimeters per year of motion across the fault; which over hundreds of years was certainly enough to build up a large earthquake. This movement is about as fast as GPS shows today across the Wasatch fault in Utah, where future earthquakes are expected. That's at least 10 times faster than what we see at New Madrid.

Because it looks like the faults that broke in 1811 and 1812 are shutting down, what will happen next? As we've seen, it would take at least 10,000 years at the current rate of motion to build up enough strain on these faults for new large earthquakes. Thus as long as GPS data don't show motion, there's probably not much to worry about on these faults.

Eventually, we expect to see strain starting to build up on one of the major faults in the area. It could be on the New Madrid fault system, another nearby fault, or one farther away like in Arkansas or the Wabash valley fault system. The Wabash is a good candidate because Mian Liu's stress calculations show that stress resulting from the 1811–1812 earthquakes could be transferred there. So far, GPS data don't show strain accumulating anywhere in the area, although there's a possible hint in data being acquired in the Wabash valley by Michael Hamburger of Indiana University.

We don't know whether significant strain accumulation will start becoming visible in hundreds, thousands, or tens of thousands of years. Once it's visible, my sense is that there would be hundreds of years of advance warning before a major earthquake, but it's important to find out.

In the meantime, the small to moderate earthquakes in the mid-continent, many of which are probably aftershocks of the 1811–1812 earthquakes, should continue. Based on experience to date, these are unlikely to be a serious problem unless one of the larger ones happens very close to a populated area.

Of course, we can never be certain because the earth can surprise us. We understand enough about the earth that it usually doesn't but not so much that it never does. Still, it makes sense to use what we do know to make the best decisions given our present knowledge.

Chapter 14

More Dangerous than California?

> Apocalyptic claims do not have a good track record. And arguments that statistics support such claims—particularly arguments that simple, easily understood numbers are proof that the future holds complex, civilization-threatening changes—deserve the most careful inspection.
>
> —Joel Best, *More Damned Lies and Statistics: How Numbers Confuse Public Issues*

The paper my coauthors and I published in 1999 presenting the GPS results showing no motion ended by stating, "It seems that the hazard from great earthquakes in the New Madrid zone has been significantly overestimated. Hence predicted ground motions used in building design there, such as the National Seismic Hazard Maps which presently show the seismic hazard there exceeding that in California, should be reduced."

This conclusion seemed obvious. Many geologists outside the USGS thought it made sense because they were skeptical of the hazard maps' claims that New Madrid was more dangerous than California. Even within the USGS, there were doubts. Although it seemed reasonable that there was some earthquake hazard in New Madrid, the idea that it was greater than California's seemed unlikely and would require a really compelling argument. It seemed unlikely even before the GPS data and seems even less likely now that these data show the ground isn't moving.

The view outside the geological community was different. Geologists develop good instincts about how the earth works that usually—although not always—point toward the right answer. It's like the way sports fans have a good idea what to expect: Northwestern might win next year's Rose Bowl (it happened in 1949), but it's not likely. However, many non-geologists didn't have enough of these instincts to be skeptical of the idea that New Madrid was more dangerous as California.

I didn't appreciate this difference until Joseph Tomasello, a structural engineer from Memphis, contacted me right after our GPS paper came out. He explained that the hazard maps were the basis for a new national building code that FEMA was promoting. Based on the hazard maps, the code required that buildings in the New Madrid area be built to earthquake resistant standards like those in California. FEMA was pressuring states in the area to write this national code into their state codes.

Joe was skeptical of what he called "the sky is falling claims" and he and some colleagues were opposed to the new code. Although its tough standards would generate a lot of business for local engineers, he thought it would be a waste of money to build to these levels. Joe and I warmed to each other quickly. I admired his willingness to stand up to FEMA and its allies. After all, he wasn't a big name in the national engineering establishment. Still, as a working engineer, he was one of the few people in the debate who had actually designed buildings. Eventually, I learned that he'd been a young Marine in Vietnam and had become pretty tough and independent.

He was pleased to hear that I was very skeptical of the hazard maps, so I offered to figure out how the mapmakers came to their surprising conclusion and how one could do better.

What Makes Sense?

Scientists typically tackle complicated questions like these using simple analyses. We look at the fundamentals of the problem and do an approximate estimate. This lets us see about what the answer should be or whether the answer from a complicated calculation makes sense. These "back-of-the-envelope" estimates often save us from embarrassing errors. They're sometimes called "Fermi problems," after physicist Enrico Fermi's exam questions like "about how many piano tuners are there in Chicago?" Fermi, who led the team that first accomplished a controlled nuclear reaction during World War II, was a master of this kind of analysis. Werner Heisenberg, Fermi's rival who led the Nazi atomic bomb effort, didn't have that talent. Heisenberg calculated that an atomic bomb would need much more uranium than it really did, which slowed the Germans greatly.

Back-of-the-envelope estimates use simple numbers, often powers of 10. The idea is to be reasonable, rather than exact. For example, I ask students to estimate about how much Americans spend each year on Halloween. They start by estimating the number of households. Because there are about 300 million people, there are about 100 million households. That makes sense; 10 million would be too few, and 1 billion would be too many. Similarly, the class estimates that each household spends about $100; $10 is too little, and $1,000 is way too high. Together, these give an estimate of about $10 billion, which is close to the $8 billion that the candy, costume, and decoration industries estimate.

Because earthquakes happen by releasing energy stored in the ground, a simple way to compare the earthquake hazard between New Madrid and California is to start with how fast the ground moves. California earthquakes result from motion between the Pacific and North American plates, which occurs at about 50 millimeters per year. New Madrid earthquakes result from the small motions within the North American plate that are much slower than 1 millimeter per year. This means that over time, New Madrid earthquakes can release much less than 1/50 of the energy in California earthquakes. However, as discussed in Chapter 7, seismic waves travel more efficiently in the Midwest. Thus a Midwest earthquake produces shaking about as strong as a California earthquake one magnitude larger. Because earthquakes of any magnitude are 10 times more common than ones a magnitude larger, we can multiply 1/50 by 10 to estimate that the New Madrid hazard is less than about 1/5 that in California.

This analysis shows that New Madrid earthquakes can be a problem but much less than in California. It's like hurricane hazards; both Boston and Miami have hurricanes, but Boston's hazard is a lot smaller. Moreover, because the recent GPS data show that New Madrid ground motion is much less than 1 millimeter per year, the New Madrid hazard looks even smaller.

Estimating Hazards

The next question was to figure out how the hazard maps' makers came up with the high New Madrid hazard. The first point to realize is that predicting earthquake hazard is hard, and no one knows how to do it well. Cinna Lomnitz of the University of Mexico, one of the founders of these analyses, described them as playing "a game of chance of which we still don't know all the rules."

Predicting earthquake hazard is much harder than, for example, predicting the weather. It's easier to make observations in the atmosphere than inside the solid earth. Most important, because things happen much faster in the atmosphere, weather forecasters have much more experience. They can see quickly how well they forecast and improve their techniques. In contrast, because of the long times between major earthquakes, seismologists have less experience and have to wait hundreds or thousands of years to see how well they forecast. Tom Parsons of the USGS notes that weather forecasters have hundreds of yearly seasonal cycles of data to work with, whereas seismologists haven't seen a full earthquake cycle anywhere within a continent. Thus, what we're doing is like trying to predict a full year's weather after watching a week in January.

As a result, there's a lot of debate and research going on about how hazard mapping should be done. Although earthquake hazard maps appear in textbooks and the news media and are used to develop building codes that affect billions of dollars in construction, there's no way to know how realistic these maps are except to wait for a very long time. It's important that people who use the maps appreciate the issues and uncertainties involved. Fortunately, these are easy to understand without going too far into detail.

Earthquake hazard maps are somewhat like the U.S. Department of Homeland Security's color-coded threat levels that purport to predict the risk of a terrorist attack. The levels, which vary from bright red to sedate green, produce both fear and ridicule. Americans don't know whether to take them seriously because we don't know how government officials set them. Presumably they reflect a mix of U.S. politics and information of varying reliability. For example, questions arose when the alert level was raised after the 2004 Democratic Convention, at a time when it looked like Senator John Kerry might pose a serious challenge to President George Bush. Even without politics, there's no way to objectively choose a level. Trying to prove that it should be orange instead of yellow would be pointless. Thus, the threat level is what the officials choosing it want us to think it is.

Like terrorist risks, earthquake hazards can't be known or measured with much accuracy. They can be estimated—a fancy word for guess—by making many assumptions. As a result, the predicted hazard depends on what those making the hazard map choose to assume.

To see why, imagine deciding whether to buy a security system to protect your house and what's in it. Some features of your house are facts—how big it is, how old it is, what it's made of, and what you have in it. The hazard is different. It's not a fact. Instead, it's something you define, so how high it is depends on how you define it. How you define the hazard depends on how much risk you're willing to take and how much money you're willing to spend to reduce the risk. Unless you're very rich, you have to think about whether that money would be better spent on something else and look for a balance.

You first worry about fire. How big is the hazard? Nationally, the average house fire causes about $20,000 damage. That's a lot of money. Maybe you should install a sprinkler system that would stop the fire and reduce the damage cost. Because a sprinkler system would only cost $2,000, installing one seems like a great idea.

Thinking about it further, you realize there's more to it than a simple comparison between the likely damage and the cost to reduce the hazard. Fires aren't common. Instead of assuming there will be a fire, it's more realistic to try to figure out what's *likely* to happen. Although that's more complicated than assuming you know what will happen, people do it all the time because the world is uncertain. This involves probability, which is the mathematical word for chance. Because people don't know the probabilities of events, all they can do is estimate them as best they can.

To estimate how likely a fire is, you use the fact that about 1 in every 200 houses has a fire during the year. That's just an average because for any house it depends on its location, how it's made, whether the occupants smoke, and so on. Still, it's a start. Using that average, you estimate that the probability of a fire is 1/200.

To figure out your hazard, you use the expected loss, which is the probability that a fire will happen times the loss that results if one does. Because the average house fire does about $20,000 worth of damage, your expected loss in a year is 1/200 times $20,000, which is $100. Now the sprinkler system sounds like a waste of money; you'd be spending $2,000 to save $100. There's probably a better use for that money.

But, you realize, there's another issue. Instead of deciding that the hazard is the expected loss in one year, you could use a longer period. If you plan to stay in the house for 10 years, you could consider the total expected loss to be $1,000. That's a lot closer to the cost of the sprinkler system, so maybe it's a good idea.

So far, you've assumed the only hazard is fire, but you could consider other hazards. How about a plane crashing into the house? That's rare—you've never personally seen it happen or known anyone who has—but it could happen. Because such crashes are rare, it's hard to estimate how likely one is to happen or what would happen if it did. You guess (wildly) that there's a 1/100,000 chance of it every year. Still, it might destroy the whole house—figure $500,000—so the expected loss is $5 per year. Over 10 years, that's $50. That's not much, so plane crashes probably won't be a big problem. But if your family stays in the house for 2,000 years, the expected loss is $10,000.

This example shows that predictions are better for shorter times because they deal with events that happen often. Experience lets us make reasonable estimates of how likely something is to occur and what will happen if it does. Thus, the more likely something is, the better we can estimate its hazard.

On the other hand, if something happens rarely, we don't have the experience to estimate its probability and effects well. Thus, estimates of rarer events over longer times are much less certain. In addition, another problem makes hazard estimates for rare events even more uncertain. The world probably won't change much in the next 10 years, but it might be very different in 2,000 years. There might not be airplanes anymore, or they might be built so that crashes almost never occur. We could try to put these possibilities into our hazard calculations, but they're hard to estimate. As a result, we wouldn't have much faith in these estimates, and it would be a long time before we knew if they were any good.

Defining Earthquake Hazards

The house hazard example gives us a way to think about earthquake hazard maps. Like the hazard from fire, earthquake hazard isn't a physical quantity that can be measured. Instead, it's something mapmakers define and then use computer programs to predict. Thus, to decide how much to believe a hazard map, it's important to know what the mapmakers assumed and what the effects of those assumptions were.

The best way to do this is to look at how maps change based on what mapmakers assume. There are four main assumptions: how, where, when, and what? The idea that New Madrid is as dangerous as California depends on choices for each of these assumptions that are at the high-hazard end of what's reasonable.

How?

Let's look at the four assumptions, starting with most important: how? Because this choice depends on policy and political arguments, the hazard is as big or as little as mapmakers want it to be. It's like the hazard to your house; how you define the hazard controls how big it is. The definition you choose depends on how much risk you're willing to take.

To see how this works, let's do a back-of-the-envelope estimate for the New Madrid zone. We can think of earthquakes as darts thrown at the map, with the same chance of hitting anywhere. The area is about 300 miles on a side (fig. 14.1). About every 150 years a magnitude 6 earthquake hits somewhere and causes moderate shaking in an area that we'll assume is a circle with a radius of 30 miles. Within the circle, some buildings might be damaged. The circle is about 3% of the total area, so in 150 years a building in the area has a 3% chance of being shaken. Assuming that a building has a typical life of 50 years, that's 1% during its life, or 0.02% in any year. So a building has 1 chance in 5,000 of being shaken in any year (0.02% is 1/5,000).

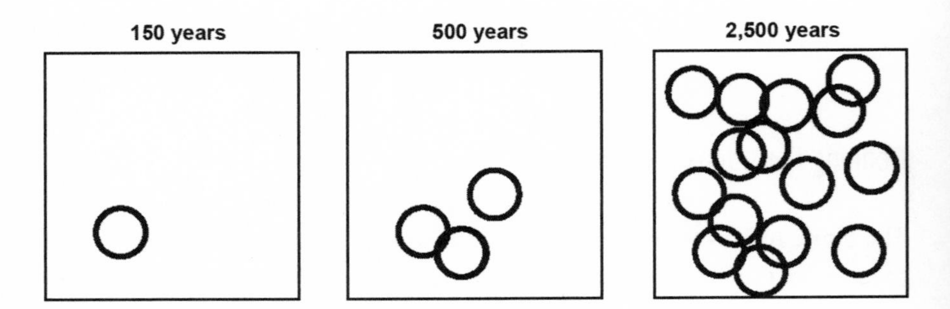

FIGURE 14.1 How earthquake hazard maps work: Assume that an earthquake of a certain size will strike within a certain time period and cause shaking within a certain area.

That's 25 times smaller than the chance of fire, which is about 1 in 200. This makes sense from our experience.

Over time, more earthquakes occur and a larger portion of the area gets shaken at least once. Some places get shaken a few times. The longer the time period the map covers, the scarier it looks. However, it's important to remember that a typical building is only there for a relatively short time, 50–100 years. If the time is 50 years, 99% of buildings will be replaced before they're ever shaken, much less damaged. If the USGS and FEMA had considered this simple fact, much of the Midwest earthquake hype could have been avoided.

Real hazard maps are just more complicated versions of this example. They include earthquakes of different magnitudes, assume that some areas are more likely to have earthquakes, and assume stronger shaking close to the epicenter. The hazard in a given location is described by the maximum shaking due to earthquakes that is predicted to happen in a given period of time.

Shaking is described by the acceleration, which is how fast the speed of ground motion changes. This is usually quoted in terms of the acceleration of gravity, called "g," which tells how fast a falling object speeds up. "g" is 9.8 meters per second, or 32 feet per second. This means that if you drop an object, it reaches a speed of 32 feet per second after one second, 64 feet per second after two, and so on.

Accelerations are what damage or destroy buildings. A house would be unharmed on a high-speed train going along a straight track because the speed isn't changing, so there's no acceleration. However, if the train stops suddenly, the house would be shaken and could be damaged if the acceleration were large enough.

As we've seen in Chapters 7 and 8, the amount of ground shaking an earthquake produces depends on its magnitude, how far away it is, and how well seismic waves travel through the ground. Two methods are used to consider these factors in predicting the hazard at a place.

One is to decide on the biggest earthquake that you're worried about. That means deciding where will it be, how big it will be, and how much shaking it will cause. This is called deterministic seismic hazard assessment.

The alternative is to consider all the possible earthquakes that could cause significant shaking at a place. This method, called probabilistic seismic hazard assessment (PSHA), involves estimating the probable shaking from the different earthquakes and producing an estimate of the combined hazard. PSHA uses the probabilities and uncertainties of factors like the locations and times of earthquakes and how much shaking will result from an earthquake of a given magnitude.

For 40 years there's been a debate about which method is better. The deterministic approach is the simplest, but makes society spend a lot of money preparing for an earthquake that's very unlikely to happen during a structure's life. Its advocates say it's especially the way to go for a critical facility such as a nuclear power plant, where a failure would be so serious that it's worth spending a lot to avoid. The probabilistic approach is closer to people's normal experience when they buy insurance or safety devices. They consider how likely something is to happen when deciding how much to invest in protecting themselves. Critics don't like PSHA because it defines the hazard in terms of a mathematical event rather than a real earthquake. In addition, its results depend greatly on the probabilities and uncertainties assumed and are very sensitive to extreme but unlikely events. This arises for the proposed nuclear waste site in Nevada, where for the long time periods being considered PSHA predicts ground motions that seem unrealistic because they're much greater than have ever been observed.

In some cases, the choice of method makes a big difference, and in others it doesn't. Both methods are often used. As probabilistic models cover longer time windows, they become about the same as deterministic ones. That's because, as figure 14.1 shows, after a long enough time any earthquake that's going to happen will have happened at least once. Moreover, many of the problems resulting from limited knowledge about earthquakes in an area apply to both methods.

Because most seismic hazard maps, including the current USGS maps, are probabilistic, let's look at that method and see the effect of different assumptions. This is important because although the ideas behind the maps are simple and sensible, they're presented in unnecessarily complicated ways. As two leaders in the field explain, the "simplicity is deeply veiled by user-hostile notation, antonymous jargon, and proprietary software." Thus, many people argue about what a hazard map predicts without understanding why it makes those predictions.

How Changing Assumptions Predicted Higher Hazard

Probabilistic hazard maps define an earthquake hazard in a given location as the maximum amount of shaking that has a certain chance of being exceeded at least once in a given period of time. These maps combine the effects of large earthquakes that are less likely to happen, but would cause more shaking, with those of small earthquakes that are likely to happen more frequently but would cause less shaking.

Around the world, earthquake hazard is most commonly defined as the maximum shaking predicted to occur at least once about every 500 years. This is the same as the maximum shaking that there's a 10% chance of exceeding at least once in 50 years. The two definitions are about the same because 50/0.1 equals 500 years. (For complicated reasons it's actually the maximum shaking expected at least once every 475 years, but 500 is close enough.)

Until recently, U.S. maps used this definition, as shown in figure 14.2. These maps predicted that the earthquake hazard in the New Madrid seismic zone was about half of that in the most dangerous places in California, because large earthquakes are rarer in the Midwest.

Things changed in the 1990s. While my coworkers and I were collecting GPS data showing that the ground in the New Madrid area wasn't moving, the USGS was ramping up its prediction of the hazard there. The USGS hazard mapping program had become more active and higher profile for several reasons. Most importantly, the USGS, FEMA, and earthquake engineers started using a new definition of the hazard. The new definition was the maximum shaking that's predicted to have a 2% chance of occurring at least once in 50 years, or on average at least once about every 2,500 years (actually 2,475).

As we've seen, the longer a hazard map's time window, the higher a hazard the map predicts. Going from 500 years to 2,500 years predicted higher hazards everywhere in the United States because it included rarer larger earthquakes. The change was largest in the east and Midwest, where big earthquakes are very rare. In fact, the new map showed the New Madrid seismic zone as more hazardous than many of California's faults (fig. 14.2).

Although the maps seem complicated, it's easy to look at one and see what it's claiming. The hazard is shown by the contour lines and shading. The

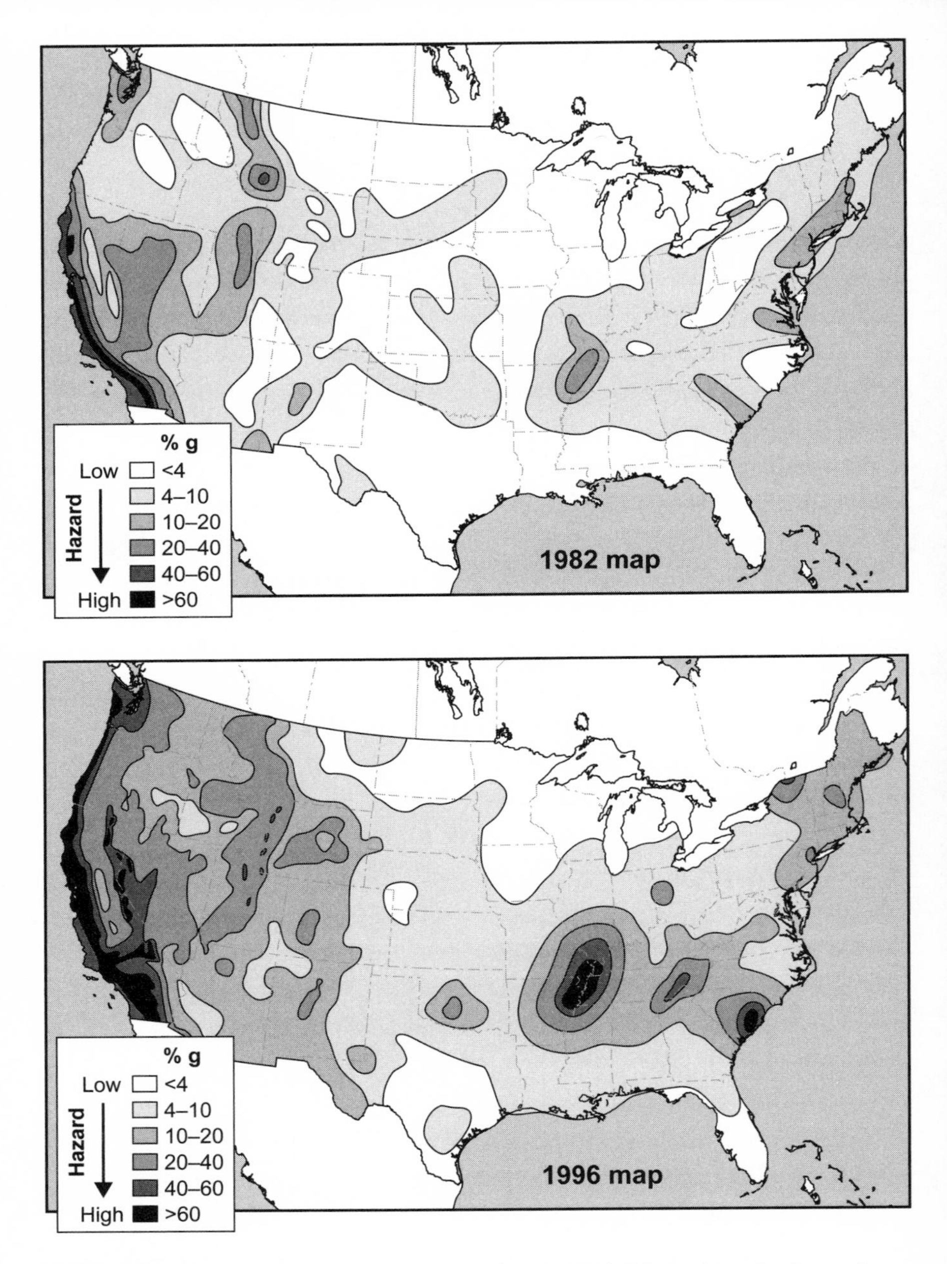

FIGURE 14.2 Comparison of the 1982 and 1996 USGS earthquake hazard maps. The predicted hazard is shown by the shading giving predicted acceleration as a percentage of the acceleration of gravity, with darker shades showing higher hazard. The new map raised the predicted hazard in the Midwest from much less than in California to even greater than California's. (After USGS)

larger numbers shown by the darker shades are higher hazard, so the predicted hazard is highest close to the most active faults.

There are easy ways to decide how much to believe maps like these. The first is to decide whether the assumptions made make sense. The second is to compare maps made using different assumptions. If changing these assumptions within the range that's reasonable given what's known and what isn't changes the map a lot, we should be skeptical about the map.

To see how changing the time window changes a map, let's look in detail at the New Madrid area (fig. 14.3). The two maps show the predicted hazard resulting from both large earthquakes on the main New Madrid faults and smaller earthquakes elsewhere. As expected, the predicted hazard is highest along the main faults and lower farther away. The only difference between the two maps is the time window.

Using the longer time window—2,500 rather than 500 years—increases the predicted hazard dramatically. To see this, look at the contour labeled 20. This corresponds to shaking that's 20% of the acceleration of gravity. Points inside this contour are approximately where some modern buildings will start to be seriously damaged, although exactly where this occurs depends on the details of each building and site geology. For the 500-year window, St. Louis is far outside this contour, and Memphis is just at the boundary. For the 2,500-year map, the predicted hazards for these cities are three and four times greater.

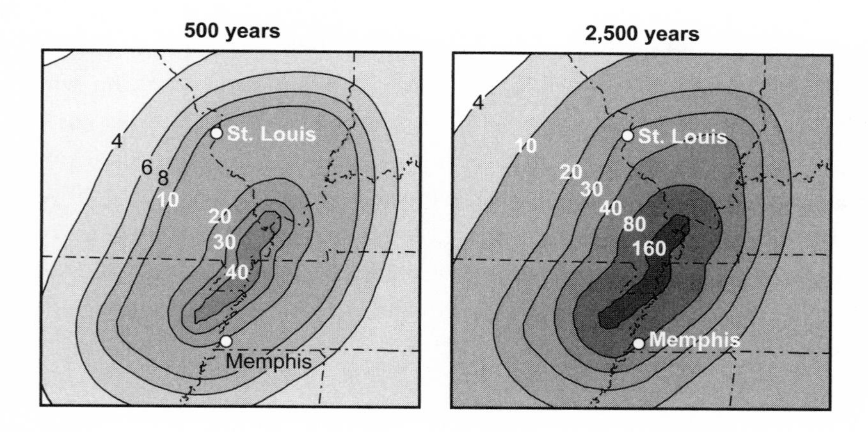

FIGURE 14.3 Comparison of the predicted earthquake hazard in the New Madrid area for different time windows. The numbers are percentages of the acceleration of gravity.

Thus, the newly predicted high hazard resulted mostly from changing assumptions, especially the time window, within a computer program. It's like deciding whether a building is in more danger from a flood or a meteorite impact. In 100 years, a flood is the greater risk. But in a billion years, both might have the same chance of happening at least once, even though floods will be much more common.

Although any hazard map's predictions have a lot of uncertainty, those for the longer time windows are much more uncertain. That's always the case because rarer events are harder to reliably predict. Thus, the assumptions in hazard maps for New Madrid are much less certain than those for California because they're based on a lot less data.

The new hazard definition had huge implications. It causes major problems for the insurance industry, and FEMA is pressuring communities in the New Madrid area to adopt building codes as stringent—and expensive—as in California. Communities are grappling with what to do.

The question is whether the "at least once in 2,500 years" criterion makes sense because it's much longer than the average life of buildings, and building to high hazard codes increases construction costs. It's not based on science but on political and policy arguments that weren't well thought out. In particular, it was chosen without considering whether the benefits are worth the increased costs.

Whether they are depends on how likely a typical building is to be severely shaken during its life, which is on average between 50 and 100 years. Designing a building for shaking that it has a 50/500, or 10%, chance of experiencing makes sense. Whether it's worth planning for shaking with a 50/2500, or 2%, chance is less clear. Given how much more this costs, it probably only makes sense for especially critical facilities like nuclear power plants. Designing for an earthquake expected on average once in 2,500 years is a much tougher criterion than used for floods, where communities typically prepare for the largest flood expected on average once every 100 years. It's also worth noting that most other countries use a 500-year criterion for earthquakes.

Where?

Once a hazard mapper decides how to define the hazard, the remaining questions are geological rather than political. Let's look at them in the order of their importance. The next question is *where* to assume that earthquakes will happen. That's tough to answer. The simplest—but often not the best—answer is to say that earthquakes will happen where they are known to have happened.

In places like California, this is a good if not perfect assumption. Seismologists don't expect major surprises on the San Andreas fault. Because the fault is slipping quickly by geological standards—about 36 millimeters per year—the largest earthquakes on each segment of the fault occur about every 200 years. Historical and geological records of shaking go back that far, so researchers think they've seen most of what's likely to happen.

Off the San Andreas, however, we've been surprised. Clarence Allan, a pioneer in studying California's faults, told our graduate school class that he used to say, "All earthquakes in California occur on mapped faults." After the 1971 San Fernando earthquake happened on a fault that wasn't on the fault map, he added "or faults that should have been mapped." Because the San Fernando fault slips much more slowly than the San Andreas—about 5 millimeters per year—it had been missed by both earthquake and geologic studies. For similar reasons, the hazard from faults buried beneath the sediments of the Los Angeles basin wasn't appreciated until earthquakes like the 1994 Northridge earthquake made it clear. Cases like these spurred programs to map faults, look for paleoseismic evidence of past earthquakes, and use GPS to find where strain is building for future earthquakes.

Places like New Madrid are even more complicated. Because the fault motion is much slower, large earthquakes are even less common, so the available earthquake history is too short to show the real pattern. As a result, hazard maps that are based only on the short earthquake history can be misleading.

One of my favorite examples is along the coast of North Africa, where the hazard map shows the highest hazard around the site of a big earthquake in 1980. Since 1999, when the map was made, the largest earthquakes have been where the map shows lower hazard. When I showed this in a department seminar, Emile Okal pointed out with impeccable French logic that "the only thing you learn from this map is the year it was made."

This short-history problem is starting to be addressed. For example, along the east coast of Canada, older hazard maps showed concentrated circles of high hazards at the sites of the big 1929 Grand Banks and 1933 Baffin Bay earthquakes discussed in Chapter 9. I got interested in these earthquakes while I was a postdoc because I thought that they happened on the faults left over from when continents split up more than 50 million years ago. In a paper, my coauthors and I pointed out that although we've seen earthquakes only on some of the faults, they're likely to happen on others. We thought that they could happen anywhere along the coast if some process made the old faults move. This seems to happen most on the parts of the coast that were covered by ice age glaciers (roughly as far south as Long Island) because removal of the ice caused stresses in the crust. Eventually—25 years later—this concept made it into the new Canadian hazard map, which "smears" out the predicted hazard along the coast. I think this makes sense, and it would be a good idea for the U.S. map.

Laura Swafford, a Ph.D. student working with me, did a simple analysis of this issue in her thesis. She assumed that the earthquake hazard was uniform along the Canadian coast and showed that about 10,000 years of earthquake history would be needed to realize it. Until then, the random pattern of where earthquakes happen would yield places with many earthquakes that looked more dangerous and places without earthquakes that looked safe. Because there's only a few hundred years of earthquake data, the earthquake history alone probably doesn't show where large earthquakes might happen on a longer time scale. To do better, we need to use the geology to identify faults that don't have many earthquakes today but might be active soon. Increasingly, hazard maps in other areas like central Europe are being changed to include the geology.

For New Madrid, the USGS maps assume that large earthquakes will occur only on the faults that are thought to have broken in 1811 and 1812, although they allow for smaller earthquakes over a wider area. What we're learning about migrating earthquakes shows that this isn't a good assumption. The current maps overpredict the hazard on the 1811–1812 faults and might underpredict it in other places. This is especially true because many of the small earthquakes we see today are probably aftershocks of large past events and so don't necessarily show where large future earthquakes will happen. When I mentioned the "Whack-a-mole" analogy to Mian Liu, he pointed out that the Chinese have a similar analogy about chasing a rabbit. Waiting for the rabbit to come out of the hole that it went down is likely to lead to a surprise because it can come out of many other holes.

When?

After assuming where earthquakes will occur, mappers have to assume *when* they will happen. This is hard anywhere and particularly hard for New Madrid. Different assumptions give quite different answers.

The key issue is when major earthquakes like the ones in 1811–1812 will happen. There are two basic approaches. We can assume that the probability of a major earthquake is constant with time or that it depends on how long it's been since the last one.

To see the difference, think of familiar examples. Time-independent probability is used when the fact that something happened doesn't change the probability that it will happen again. Even after flipping a coin three times and getting a head each time, the chance of getting a head on the next flip is still 50%. On the other hand, in card games the probabilities are time-dependent. After two aces have been drawn, the probability that the next card will be an ace goes down.

For earthquakes, the simplest assumption is that the probability of a major earthquake is time-independent. If so, a future one is equally likely to happen 10 days, 10 years, 100 years, or 1,000 years after one occurs. Thus, if major earthquakes occur on average 500 years apart, the chance of having one in the next 50 years is 50/500, or 10%. With this time-independent model, which is what the USGS uses for New Madrid, an earthquake can't ever be overdue.

This is also how researchers describe floods. Although the "100-year" flood is the biggest one expected on average in 100 years, having this flood one year doesn't reduce the chance that another will happen next year. Similarly, the fact that an area was hit by a hurricane doesn't seem to affect the chance that it will be hit again in the next year.

Earthquakes are different because in the earthquake cycle strain builds up slowly after one big earthquake until there's enough strain for another. This cycle is described using a time-dependent model, where the probability of a large earthquake is small shortly after one happens and then increases with time. This assumes that the recurrence time between large earthquakes has a mean and standard deviation described by a probability curve, which can be a bell curve or some other. For example, we can use a bell curve with a mean of 500 years and a standard deviation of 100 years. In this case, the most likely time between earthquakes is 500 years, and there's a 95% chance that this time will be longer than 300 years and less than 700 years.

Because the probability changes with time, a time-dependent model predicts that an area has a lower probability of an earthquake—and thus a lesser hazard—if less than about two-thirds of the assumed average time between earthquakes has passed. However, if a large earthquake hasn't occurred by this two-thirds time, a time-dependent model will predict higher probabilities and thus a higher hazard than the time-independent model. Past this time, an earthquake can be considered "overdue."

If major New Madrid earthquakes occur on average 500 years apart, this "crossover" time is the year 2144, which is two-thirds of 500, or 333 years after 1811. Figure 14.4 shows that until about the year 2150, a time-dependent model predicts a lower probability of a major earthquake than a time-independent model. The exact numbers depend on the probability curve that is assumed to describe the recurrence times. Using a bell curve with a mean of 500 years and a standard deviation of 100 years, the time-dependent model predicts that the probability of a major earthquake is less than 1%. This is 10 times smaller than that from the time-independent model that the USGS and FEMA use.

These earthquake probabilities cause a lot of confusion for two reasons. The first is that people talk about "the" probability of a large earthquake in the New Madrid zone, as though this were a fact that scientists know. In reality, the probability isn't something we know or even can know. All we can do is estimate it by choosing a model, a probability curve, and the numbers that describe that curve. Depending on what we assume, we can easily get any number between 0 and 10% as well as larger values with only a little effort. It's important to say which assumptions are made in computing the probability being discussed.

The second source of confusion is that people often don't say what magnitude earthquake they're talking about. This makes a huge difference. Using a time-independent model, the probability in the next 10 years of a major magnitude 7 earthquake is 10/500, or 2%, assuming that one happens about every 500 years. That's very different from the 10/175, or 6%, probability of a magnitude 6, which happens somewhere in the zone about every 175 years, and the 10/120, or 50%, probability of a magnitude 5, which happens about every 20 years. Confusing these is like comparing the probabilities of a hurricane, a rainstorm, and scattered showers.

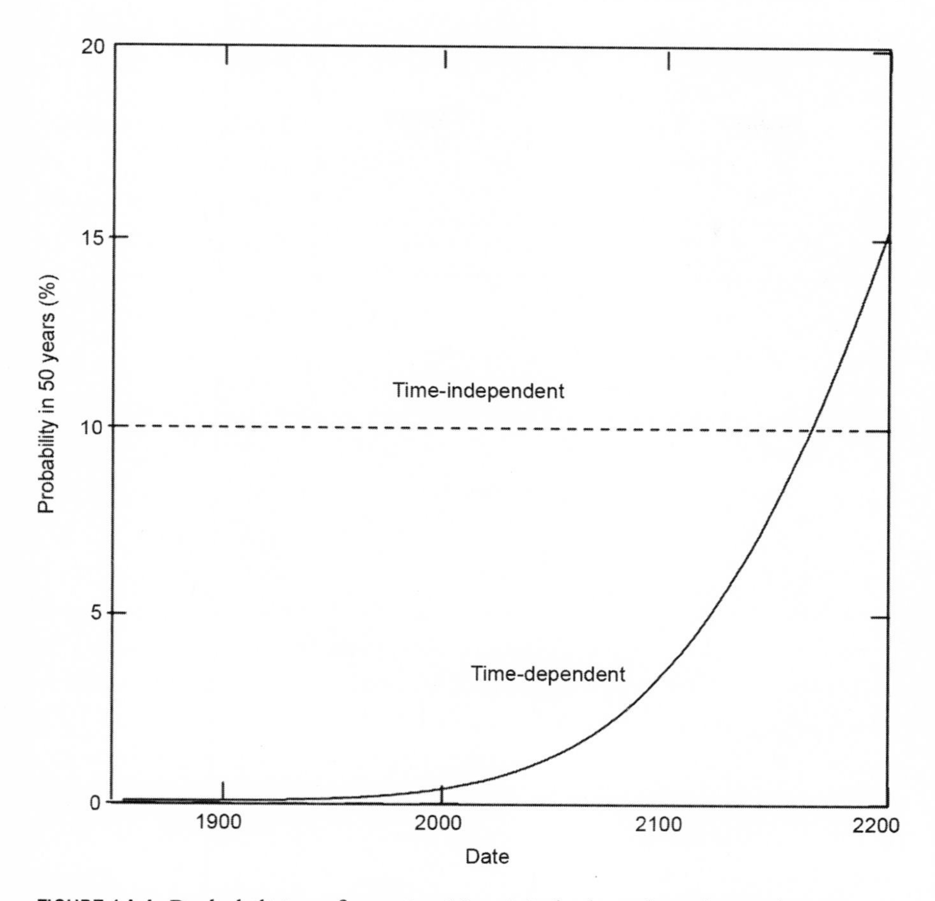

FIGURE 14.4 Probabilities of a major New Madrid earthquake in the next 50 years as a function of time since 1812 computed for different models. (Stein et al., 2003)

Which model predicts higher hazard depends on how far a fault is into its earthquake cycle. That depends on how long it's been since the last big earthquake and on the average time between big earthquakes. To see this, let's look at a few seismic zones (fig. 14.5). For New Madrid and Charleston, time-dependent models predict lower probabilities of large earthquakes, and thus lesser hazards than time-independent models, because both zones are early in their cycles. Charleston is very early in its cycle because the time since the large earthquake in 1886 is only 22% of the 550-year cycle assumed from paleoseismology.

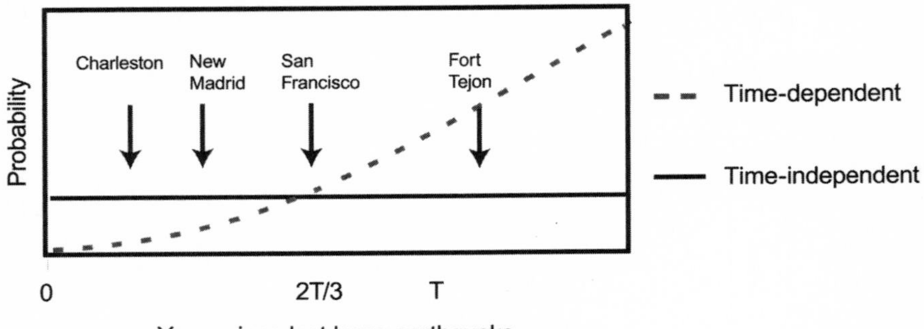

FIGURE 14.5 Comparison of probabilities of a large earthquake predicted by the two models for different faults. T is the average time between large earthquakes. (Hebden and Stein, 2009)

At New Madrid, it's been 200 years since 1812, which is 40% of the assumed 500-year cycle. James Hebden, a Northwestern undergraduate, calculated in his thesis that the time-dependent model predicts hazard in Memphis for the years 2000–2050 that's about one-third less than predicted by the time-independent model. I was impressed by how interested he was and the fine job he did, and couldn't tell why until at graduation I learned that his father was an executive with a major insurance company, which has to think about earthquake risks.

Which model is better? At this point, no one knows. Part of the problem is short earthquake records, but even where there are long records it's not clear. Different studies have argued for using one model over the other. It's an important question because time-dependent models predict that places where large earthquakes have happened recently are less, not more, dangerous. Moreover, even once we've decided on which kind of model to use, we can choose the kind of probability curve to use and the numbers that describe it.

The USGS is now using time-dependent models for the southern San Andreas fault that predict higher hazard than their earlier models. That's because it's been 152 years since this fault segment last broke in the Fort Tejon earthquake, which is longer than the 132-year average recurrence coming from the earthquake history shown in Chapter 9. I think this makes sense.

However, if USGS uses time-dependent models for California, it should be consistent and also use them for New Madrid and Charleston, which would lower the predicted hazard from their current maps.

There's a further complication for New Madrid because GPS measurements don't show the ground motion that would be expected if energy were being stored before an upcoming earthquake. The fault system looks like it's shutting down for a long time, so the hazard is probably much less than either model predicts. Hopefully in the future, we'll learn more about what's going on and put this information to good use.

What?

The last question is *what* to assume will happen when a large earthquake occurs. This involves two assumptions: how large these earthquakes will be and how much shaking they will produce. In places like California, where there are seismograms from many large earthquakes, reasonable assumptions can be made based on good data.

For New Madrid, hazard mappers assume that the biggest earthquakes will be like those that happened in 1811 and 1812 and that they will produce similar shaking. Each of these assumptions introduces a lot of uncertainty. As discussed in Chapter 5, there's a major debate about the magnitude of the 1811–1812 earthquakes because the seismometer hadn't been invented. For the same reason, it's hard to put numerical values on how much shaking occurred from the written descriptions of shaking and damage. These uncertainties combine to make the hazard even more uncertain.

Andy Newman and I wanted to look into the effect of the assumptions that the USGS mappers used for New Madrid but couldn't get their computer programs. Fortunately, we knew researchers at a company in Chicago called Impact Forecasting that did similar studies for Aon Insurance. Andres Mendez and John Schneider, who has since become head of risk analysis for Australia's geological survey, had similar programs. This let us do what the high-tech industry calls "reverse engineering": figuring out how a proprietary technology works. Andy set up these programs so that making the same assumptions as the USGS gave the same maps.

We knew that what the mappers assumed for magnitude had a major effect on the maps. What surprised us was how much their assumptions about shaking also affected the maps. These assumptions are made by choosing a ground motion attenuation model, which is a mathematical formula saying how large the shaking will be at a certain distance from an earthquake of a certain magnitude. In areas like California, these models are based on a lot of data. However, because large earthquakes are rare in the Midwest, we don't

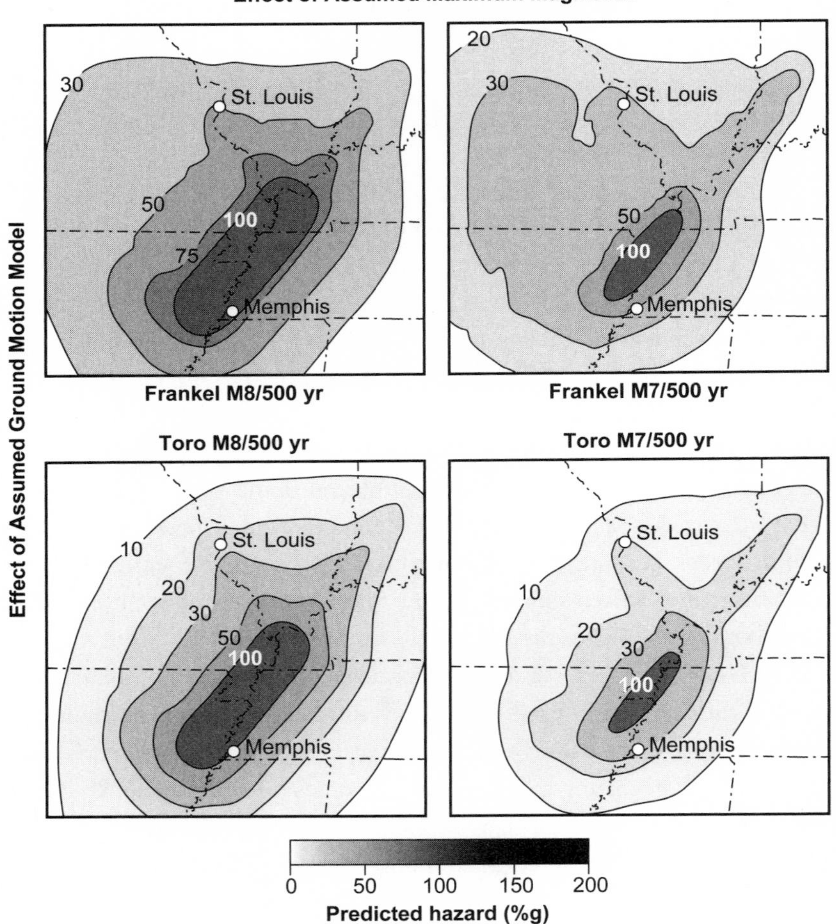

FIGURE 14.6 Comparison of seismic hazard maps made assuming different magnitudes of the largest New Madrid earthquake (rows) and different ground motion models (columns). (Newman et al., 2001)

have seismograms from large earthquakes. Instead, researchers combine data from small earthquakes with theoretical results. Not surprisingly, this produces different models that predict very different amounts of shaking.

Digging into a USGS report, we were intrigued to read that those making the maps "were not comfortable" with the model that had been used before. Instead, they "decided to construct" a new model, which predicted much higher ground motions.

To see the effects of the mappers' choices, Andy made maps using different assumptions. Figure 14.6 shows four possible maps. The two in each row are for the same ground motion model but have different values of the "maximum magnitude," or the magnitude of the largest earthquake. The maps use a time-independent model in which this earthquake is predicted to occur on the main faults on average every 500 years regardless of how long it's been since the past ones in 1811–1812. Raising this magnitude from 7 to 8 increases the predicted hazard at St. Louis by about 35%. For Memphis, which is closer to the main faults, the predicted hazard more than doubles. Thus the assumed maximum magnitude of the largest earthquake on the main faults affects the predicted hazard most near those faults.

The two maps in each column have the same maximum magnitude but different ground motion models. The model developed by Arthur Frankel of the USGS predicts a hazard in St. Louis that is about 80% higher than predicted by a model developed by Gabriel Toro of Risk Engineering, Inc. For Memphis, this increase is about 30%. The ground motion model affects the predicted hazard all over the area because shaking results from both the largest earthquakes and from smaller earthquakes off the main fault.

The four maps together show that although there's a lot of argument about the maximum magnitude to use in a hazard map, the choice of ground motion model is at least if not more important.

How Much Do You Want It to Be?

Comparing the different hazard maps is sobering. It shows that, depending on the assumptions, a mapmaker can produce a wide range of maps predicting very different levels of hazard. That's just the way it is, not a criticism of hazard mappers. It's like the joke that if you ask, "How much is 2 + 2?" a geophysicist will say, "How much do you want it to be?"

Given this problem, hazard modelers use a "logic tree" to combine different options. For example, figure 14.7 shows how the four models shown in figure 14.6 can be combined. Starting on the left, there are two different magnitudes for the largest earthquake. For each, there are two different ground motion models. This gives four "branches" of the tree. Each branch has a different prediction for the level of shaking at any point on the map, as shown by the values for St. Louis.

These predictions are combined by assigning probabilities to each branch of the tree. For illustration, let's give a 70% probability to magnitude 7 and 30% to magnitude 8. This is about the same as saying that the magnitude is $(0.70 \times 7) + (0.30 \times 8) = 7.3$. Next, let's give an 80% probability to the Toro ground motion model and 20% to the Frankel model. We then multiply the numbers on each branch. For example, the top branch gives $0.70 \times 0.80 \times 0.20 = 0.11$ g, or 11% of the acceleration of gravity. Adding up all four branches gives a value of 0.26 g, or 26% of the acceleration of gravity. This is the predicted value for the shaking at St. Louis coming

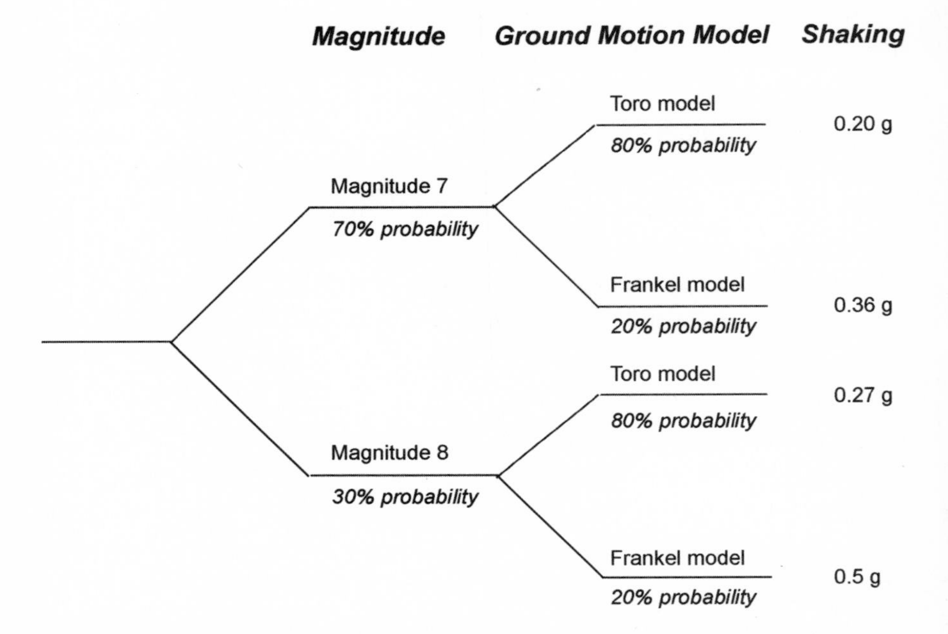

FIGURE 14.7 Logic tree combining the four models in figure 14.6 to predict shaking at St. Louis.

from a weighted combination of the four different predictions. This value is somewhere between the four predictions and is most influenced by the branches to which we assigned the highest probability.

An actual hazard model is like this, only with more and longer branches. It can include more than two choices for each parameter as well as choices for the locations of earthquakes and for when they'll happen.

Logic trees are a way to include alternative predictions, but still have a major problem. The results depend on the assumptions put into the tree and the probability given to each one. These in turn depend on what the mapmakers think is happening today and will happen in the future. For New Madrid, these questions aren't resolved and probably won't be for hundreds or thousands of years. There's no objective way to decide what to assume for the different choices that have to be made. Because the process is very subjective, the mapped hazard depends on what the mapmakers think the hazard is.

As we've seen, the USGS chose assumptions on the high side of the range of values that seem reasonable. These come from their belief that there's a major hazard because major earthquakes have happened, so another is coming soon. Thus, although they say that the hazard maps prove that New Madrid is very dangerous, it's really that the maps are made *assuming* that New Madrid is very dangerous. For example, the mappers considered but didn't use a model in which big earthquakes are about 1,500 years apart, which they admitted "produced substantially lower probabilistic ground motions."

Even now, the USGS maps don't use the newest and best data available. These are the almost 20 years of GPS results that show no sign of a major earthquake on the way. Including the GPS data would lower the predicted hazard dramatically. I think a more sensible map would use the GPS data. For example, we could assume that the biggest earthquake expected would be magnitude 6.7, which is about the largest aftershock we'd expect from the 1811–1812 earthquakes. We would also use the 500-year time window.

As figure 14.8 shows, these assumptions predict a hazard five to ten times less than the USGS map. In this scenario, the earthquake hazard is there but modest. That's a lot less scary than the USGS map, but I think it's more realistic given what's now known. There's also the option of using a logic tree that gives equal weight to the GPS data and the older assumption that a big earthquake is coming, which would give a result in between.

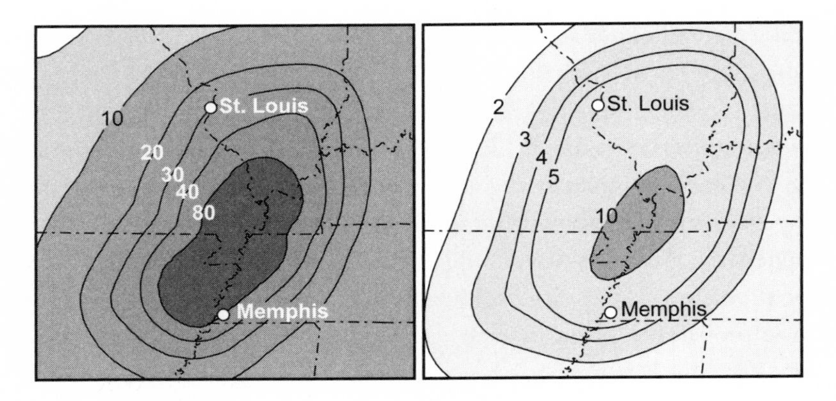

FIGURE 14.8 *Left,* Seismic hazard map made using USGS assumptions; *Right,* Seismic hazard map based on the GPS data and a 500-year window. The numbers are percentages of the acceleration of gravity.

Caveat Emptor

The best approach to earthquake hazard maps is the Latin phrase "caveat emptor," which means "let the buyer beware." Hazard map predictions vary in quality. Those for California are fairly robust, which means that using values in the range of what seismologists think they know give similar predictions. Thus, the predictions are probably pretty good. For New Madrid, wildly different predictions can be made using different assumptions about what's going on. As a result, the predictions are much more likely to be wrong. The best bet is to look at a map, ask whether the predictions make sense, and act accordingly. As the old adage says, statistics like these should be used the way a drunk uses a lamppost—for support rather than illumination.

It's interesting to compare hazard maps to other models that try to predict what nature will do. Weather forecasts rely heavily on observations of weather somewhere else that's known to be on the way. Thus, they're pretty good at predicting how a system that is mostly understood will behave in the next few days. However, problems like those for earthquake hazards come up in other situations that also use inadequate data for long-term forecasts of complicated systems that aren't well understood. The models can be useful, but they can also cause difficulties, especially when the government uses them to make policies.

These challenges are recognized, and there's a lot of thinking going on about them. For example, in scientific historian Naomi Oreskes' analysis:

> Forecasts of events in the far future, or of rare events in the near future, are of scant value in generating scientific knowledge or testing existing scientific belief. They tell us very little about the legitimacy of the knowledge that generated them. Although scientists may be enthusiastic about generating such predictions, this in no way demonstrates their intellectual worth. There can be substantial social rewards for producing temporal predictions. This does not make such predictions bad, but it does make them a different sort of thing. If the value of predictions is primarily political or social rather than epistemic, then we may need to be excruciatingly explicit about the uncertainties in the theory or model that produced them, and acutely alert to the ways in which political pressures may influence us to falsely or selectively portray those uncertainties.
>
> As individuals, most of us intuitively understand uncertainty in minor matters. We don't expect weather forecasts to be perfect, and we know that friends are often late. But, ironically, we may fail to extend our intuitive skepticism to truly important matters. As a society, we seem to have an increasing expectation of accurate predictions about major social and environmental issues, like global warming or the time and place of the next major hurricane. But the bigger the prediction, the more ambitious it is in time, space, or the complexity of the system involved, the more opportunities there are for it to be wrong. If there is a general claim to be made here, it may be this: the more important the prediction, the more likely it is to be wrong.

Similarly, Orrin Pilkey, a leading coastal geologist, and environmental scientist Linda Pilkey-Jarvis conclude in their book *Useless Arithmetic: Why Environmental Scientists Can't Predict the Future:*

> The reliance on mathematical models has done tangible damage to our society in many ways. Bureaucrats who don't understand the limitations of modeled predictions often use them . . . Agencies that depend on project approvals for their very survival (such as the U.S. Army Corps of Engineers) can and frequently do find ways to adjust the model to come up with correct answers that will ensure project funding. Most damaging of all is the unquestioning acceptance of the models by the public because they are assured that the modeled predictions are the way to go.

Chapter 15

Chemotherapy for a Cold

Don't rush to get the wrong answer as fast as possible.

—Computer programming adage

The legal concept "fruit of the poisonous tree" says that if police acquire evidence illegally, for example by tapping telephones without a warrant, then neither that evidence—the tree—nor any fruit resulting from it can be used in court. In the same way, because earthquake hazard maps for New Madrid are developed by assuming that a major earthquake is on the way, everything based on these maps has problems.

Questions should have arisen when the 1996 hazard maps were being made. If I'd told a seminar audience in my department that new results showed that Memphis was more dangerous than California, I'd have gotten a lot of skeptical questions. Within a few minutes the assumptions leading to that conclusion would have been under intense scrutiny. However in the insular world of the government's hazard modeling, hard questions weren't asked and the surprising result went unchallenged.

These questionable assumptions propagated into the building codes that FEMA pressures communities to adopt. This occurs by a complicated process through a FEMA-sponsored organization of earthquake engineers called the Building Seismic Safety Commission. Hazard maps are turned into design maps specifying the levels of shaking that buildings in different areas should withstand. These results are incorporated in codes specifying how buildings should be constructed to achieve this goal.

Usually, this works well. Starting with a sensible hazard map, sensible building codes come out. In California, codes that evolved over many years based on a long earthquake history and experience from many damaging earthquakes have ensured that few buildings collapse. The additional cost of constructing earthquake-resistant buildings in this earthquake-prone area seems to be in reasonable balance with the benefits.

However, the new USGS maps produced a big problem. Changing the predicted ground motion from the maximum expected at least once in about 500 years to at least once in about 2,500 years made the predicted hazard much greater everywhere, not just in the New Madrid area. In particular, it became much higher in California.

This wasn't acceptable to California, which is the national center of earthquake engineering. The state's universities, private industry, and government have considerable experience with developing and working with earthquake building codes. They immediately realized that the new maps predicted much stronger shaking than was thought to have happened in past earthquakes, and that building to withstand this would greatly increase construction costs. The answer was to "cap" the predicted ground motion at a level that seemed reasonable but was much lower than the new maps predicted. This amounted to staying with what already worked.

A red flag should have gone up. If the new maps didn't work in California, where much is known about earthquakes and there's considerable experience incorporating earthquake resistance into building codes, why use them in the Midwest? The Midwest doesn't have this experience precisely because damaging earthquakes are so much rarer. If they're so much rarer, how could it make sense to build to the same levels as in California? Wouldn't that cost more than it accomplished?

Asking these questions would have shown the problems with the new maps, but they weren't asked. That's because most midwestern states didn't have the experience to ask. If they had, less stringent and less expensive building codes would probably have been proposed. The GPS results would have been a neat scientific discovery but with far fewer implications for society.

Keeping Buildings Standing

The goal of earthquake engineering is to keep structures from collapsing due to the strongest shaking that they might be expected to encounter and to minimize the damage from weaker shaking. Structures, by the way, are anything that people build; buildings are structures that people spend a lot of time in.

People have been building things for thousands of years, and they rarely fall down. That's because engineers and architects have a lot of experience and do careful design. Making structures stand up in earthquakes is harder and takes more work. There's been considerable progress in the past years, and research continues about how to do better.

Under normal conditions, a building is designed to support its own weight, plus the weight of what's inside: people, furniture, hot tubs, library books, etc. This weight is a static load that doesn't change much over time. It pushes down and compresses the building material, so the building is made of a material that is strong enough in compression to hold the building up. These materials include adobe (mud brick), stone, brick, wood, concrete, and steel.

Staying up in an earthquake is different. The moving ground applies a dynamic load, one that changes quickly. To see what happens, stand a book on a piece of paper and pull the paper to the left. As shown in figure 15.1, the book falls to the right. That's because of Newton's famous first law of motion: An object at rest tends to stay at rest. This tendency is called inertia. The book "tries" to stay up, so as its bottom moves left, an inertia force moves the book the other way relative to the paper.

Now, imagine what happens as a brick building is shaken (fig. 15.2). When the ground moves one way, the building tries to move the other way. These

FIGURE 15.1 When a book's bottom is moved in one direction, inertia makes it fall in the opposite direction.

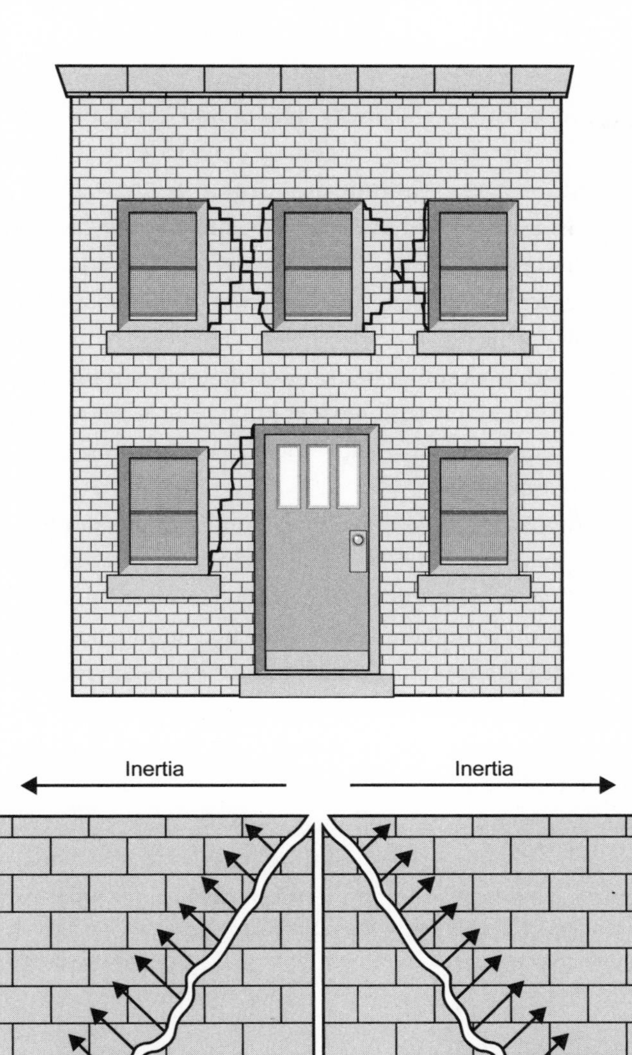

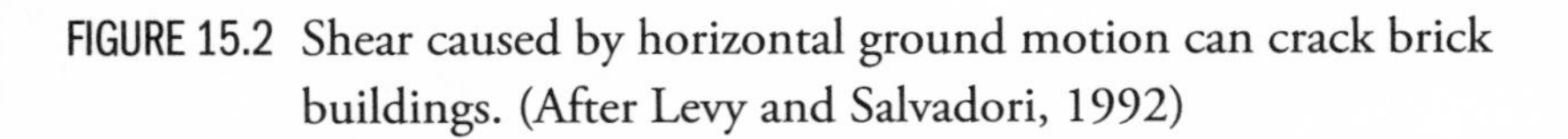

FIGURE 15.2 Shear caused by horizontal ground motion can crack brick buildings. (After Levy and Salvadori, 1992)

forces combine to distort, or shear, the building. Each row of bricks tries to slide over the one below, so the bricks start to pull apart. Because only weak cement mortar holds the bricks together, cracks develop. As the ground moves back and forth, the building sways, and the cracks widen and connect. Brick buildings damaged by an earthquake, like the ones that survived the 1886 Charleston earthquake, often have X-shaped cracks between windows. If the shaking is strong enough and goes on long enough, the building can crumble and collapse.

The problem is that although brick construction is strong in compression, when squeezed, it's weak in tension when pulled apart. Many building materials have this difference. This makes them good for most applications but vulnerable in earthquakes.

How much damage results from a given amount of shaking depends a lot on the building material. The most vulnerable common material is adobe. In December 2003, a magnitude 6.6 earthquake struck the city of Bam in Iran. Although the earthquake wasn't that big, more than 25,000 people died, in large part because many adobe buildings collapsed, including a famous ancient citadel.

Brick or concrete block constructions, which are both called masonry, are better than adobe but still vulnerable. This is an issue in places like the Midwest that have many brick buildings. New masonry construction can be made stronger by including steel reinforcing bars in the brick walls. This works because steel is very strong in tension; steel cables hold up suspension bridges. There are also ways to strengthen, or "retrofit," existing brick buildings, but these can cost almost as much as or even more than a new building. As a result, unreinforced masonry buildings can still be found in California and other earthquake zones.

Wood is a much better material because it's strong in tension. That makes it ductile, which means it can bend without breaking. In addition, wood is very light. Thus, the forces caused by ground shaking, which are proportional to a building's weight, are smaller for wood structures than ones built of most other materials. As a result, wood houses generally do well in earthquakes. I once talked to USGS geologist George Plafker, who studied the 1960 Chile and 1964 Alaska earthquakes, which are the two biggest since the invention of the seismometer. He told me that he didn't have earthquake insurance on his wood frame house, only a few miles from part of the San Andreas fault that broke in the 1906 earthquake. Even if a big earthquake happened, he doubted that it would do enough damage for him to collect because typical earthquake insurance policies only pay for damage that is more than 15% of the house's value.

Most major structures today are made from concrete. Concrete is made by mixing aggregate materials like gravel and sand with cement that hardens once water is added. It's amazingly popular. In a year, about a cubic yard is made for every person on earth.

Concrete itself is weak in tension because the cement holding the aggregate together can crack. However, in the 1800s engineers started using reinforced concrete with steel reinforcing bars, called "rebar," inside. Reinforced concrete is very strong and makes it possible to build huge strong structures.

However, even concrete structures can be destroyed by earthquakes. To see how, let's look at a modern building (fig. 15.3). Typically, a building's weight is supported by concrete columns, and the walls are made of glass or other material that doesn't support the weight. The columns support the horizontal beams and floors, which are often concrete slabs. If earthquake shaking moves the building too far, the columns can fail. The first story collapses, often followed by the others, and the building "pancakes." This happened during the 2010 Haiti earthquake to many buildings that were made of substandard concrete or cinder blocks without adequate steel reinforcement.

The first story is especially vulnerable for several reasons. Its columns support the most weight. It's often used for car parking or another application that favors more open space, so there are fewer columns or fewer walls.

FIGURE 15.3 A reinforced concrete building can collapse if the columns fail. (After FEMA)

In addition, it's often taller than the other stories, so the columns can be deformed more.

That happened to Olive View Hospital in the 1971 San Fernando earthquake. As figure 15.4 shows, the columns in the weak first story were seriously damaged. Although the building didn't collapse, it had to be demolished. While I was a postdoc at Stanford, I teased Allan Cox, the dean of earth sciences, that his building looked like Olive View Hospital. Because his office was on the ground floor, beneath the second floor library with its heavy load of books, and he was a lot older than my 25 years, he didn't think this was as funny as I did. Allan said in his serious deanly way that he'd been assured that the building was safe, and I replied that we'd know when the 1906 earthquake repeated.

Such damage from the San Fernando earthquake showed that even concrete buildings were vulnerable if they weren't ductile enough. There are ways to make them more ductile so that they deform without collapsing. A common way is to increase the amount of steel in the concrete. In columns, the steel runs both vertically and horizontally, sometimes as a spiral around the columns. The floors and beams are strengthened to resist shearing. All parts of the building—columns, beams, floors, and walls—are strongly connected and tie the building together.

FIGURE 15.4 Damage to the first floor columns of Olive View Hospital in the 1971 San Fernando earthquake. (USGS)

Engineers also have other methods to make buildings earthquake-resistant. One is to build buildings with ductile steel frames. Another is to include shear walls, or strong walls that prevent the building from shearing too far. Caltech's Seismological Laboratory building, constructed shortly after the San Fernando earthquake, has a shear wall inside that should prevent the embarrassment of its being leveled by an earthquake. A new approach, called base isolation, is to put rubber or lead bearings between the building and its foundation so that much of the horizontal ground motion isn't transmitted into the building.

Figure 15.5 puts these ideas together. It shows the approximate percentage of buildings of different construction that collapse, depending on the intensity of earthquake ground shaking. In general, reinforced concrete buildings designed for earthquake resistance do the best, followed by ordinary reinforced concrete, wood frame, brick, and adobe.

To put this in context, think of the December 1811 New Madrid earthquake. As we saw in Chapter 5, the intensity of shaking got smaller with

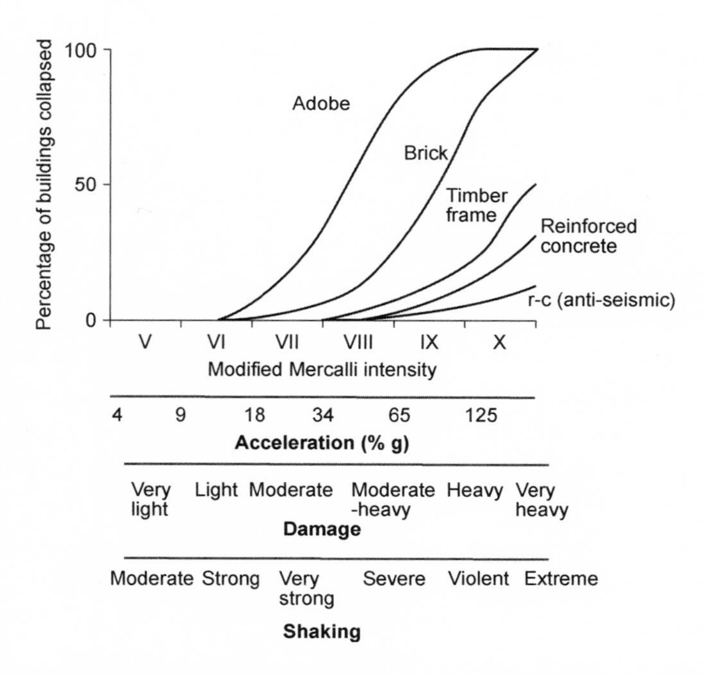

FIGURE 15.5 How vulnerable buildings are to earthquake damage depends on the material used in their construction. (Stein and Wysession, 2003)

increasing distance from the earthquake. New Madrid itself experienced shaking with intensity about IX. If there had been unreinforced brick buildings there, about half would have collapsed. About 20% of wood frame houses would have collapsed as well as 10% of reinforced concrete buildings (which hadn't been invented yet). Farther from the earthquake, Memphis (which didn't exist yet) would have experienced shaking with intensity about VII, which would have collapsed about 5% of unreinforced brick buildings, but few if any wood frame or concrete buildings. Even farther away, the shaking at St. Louis (which did exist) was intensity VI, and buildings didn't collapse.

Figure 15.5 is just an average for many reasons. How buildings fare depends on their design and quality of construction as well as the construction material. Similarly, how a building responds to shaking depends on more than the intensity, which is just a simple way to describe the shaking. The duration of strong shaking matters because the longer it goes on, the more damage it can produce.

Another important factor is the period of the seismic waves. As mentioned in Chapter 7, a building can be thought of as a spring-mass system. Like a playground swing, a building has a natural period of oscillation that it would sway back and forth with if you gave it a push. This period is about one second for every 10 stories, so a 20-story building has a natural period of about two seconds. If a building is shaken by ground motion with a period close to its natural period, the shaking is much greater. This effect, called resonance, is like pushing on a swing every time it reaches the same point in its path. Hence, buildings suffer the worst damage from ground motion with a period close to their natural periods.

For example, video taken in Banda Aceh, the closest city to the great 2004 Sumatra earthquake, shows strong ground shaking with intensity about VII. However, damage from the shaking (as opposed to the tsunami) was much less than might have been expected for such a huge earthquake. Most of the buildings, which were one or two stories high, were largely undamaged. However, taller buildings were damaged, and some collapsed. This pattern arose because the earthquake was far enough away that most of the shorter period waves that would do the most harm to low buildings had decayed while the surviving longer period waves damaged taller buildings.

The ground a building is built on also matters. As discussed in Chapter 7, seismic waves get bigger in soft material, so shaking there is stronger than on harder rock. The shaking intensities reported at many of the sites for the 1811–1812 earthquakes include this effect. Moreover, the shaking can liquefy soft ground, as we talked about in Chapter 5. In the 1964 Niigata, Japan, earthquake, buildings that survived the shaking with little damage toppled when the ground liquefied.

Tough Choices

Deciding how to prepare for earthquakes in the New Madrid region is complicated. We know three basic things. First, if a major earthquake like those in 1811–1812 happened again, it would be very destructive. Second, although building to the standards used in California would be very expensive, it would reduce the damage significantly if a major earthquake happened. Third, although we expect small earthquakes, there's no sign that a major earthquake is on the way.

The policy the federal government is pushing is "better safe than sorry." Even if a major earthquake is unlikely, why not prepare? We don't know enough to say that a major earthquake is impossible. The weakness in that argument is that it ignores the fact that earthquakes aren't the only—or biggest—problem communities face.

I sometimes discuss this issue with other Midwest geologists using a simple thought experiment. I ask them to imagine that their department is planning a new building. They discuss earthquake-resistant construction with the architect and engineer, who are happy to provide any level of seismic safety requested. However, the more safety the department wants, the more the building will cost.

As geologists, their first instinct is to make the building as safe as possible. However, they quickly realize that earthquake resistance is just one of the ways to use the money. The more of the budget they put into seismic safety, the better off they'll be if a large earthquake seriously shakes the building during its life. However, they're worse off otherwise because that money wasn't used for office and lab space, equipment, etc. After all, the odds are good that the building will never be seriously shaken.

Deciding what to do involves cost-benefit analysis. This means trying to estimate the maximum shaking expected during the building's life and the level of damage to accept if that happens. The scientists have to consider a range of scenarios, each involving a different cost for seismic safety and a different benefit in damage reduction. They have to weigh these, accepting that estimates for the future have considerable uncertainties, and somehow decide on what they think is a reasonable balance between cost and benefit.

People make similar decisions when they buy insurance for their homes, cars, or lives. The insurance company will sell them as much insurance as they want. The more they buy, the better they're covered, but the more it costs. That money can't be used for other things. Thus, instead of buying as much insurance as they can, people buy an amount that balances the costs and benefits based on their personal preferences.

This process is what communities have to do in preparing for earthquakes. The first part of the process, looking at costs and benefits, boils down to two simple principles.

The first principle is "there's no free lunch." This means that resources used for one goal aren't available for another. That's easy to see in the public sector, where there are direct trade-offs. Funds spent putting steel into school buildings can't be used to hire teachers. Money spent strengthening hospitals can't be used to treat uninsured patients. Spending on stronger bridges can mean hiring fewer police officers and firefighters.

A similar argument applies to saving lives. Stronger buildings might over time save a few lives per year. On the other hand, the same money invested in other public health or safety measures (free clinics, flu shots, defibrillators, more police, highway upgrades, etc.) would save many more lives.

The second principle is "there's no such thing as other people's money." The reason there's been little discussion of the costs of stringent construction standards is the assumption that someone else will pay. The federal government is pushing what's called an unfunded mandate: rules that someone else will pay to implement. The idea is that states and communities will, in turn, make rich developers or big businesses pay.

Unfortunately, this doesn't work. Everyone ultimately pays. That's easy to see for public buildings, which will cost taxpayers more. It's also true for the private sector. The added costs for safer buildings affect the whole community. Some firms won't build a new building or will build it somewhere else. This reduces economic activity and could cause job losses. In turn, state and local governments have less tax money for community needs: schools, parks, police, etc. For those buildings that are built, the added costs get passed on directly to the people who use the building—tenants, patients, clients, etc.—and then indirectly to the whole community.

As a result, mitigating the risks to society from earthquakes (or other natural disasters) involves complicated economic and policy issues. Seismologists and engineers tend to focus on the scientific problem of estimating the hazard and the engineering task of designing safe structures. These are tough but technical. The harder challenge is to develop sensible policies that balance costs and benefits, given what's known and isn't known about future earthquakes and their effects.

These choices are hard even in California, where the earthquake hazard is high and there's a lot of experience dealing with earthquakes. For example, after hospital buildings collapsed in the 1971 San Fernando earthquake, the state required that hospitals be strengthened. Hospitals, like schools, have the special problem of long lifetimes. That's because the communities or nonprofit organizations that run them often don't have a capital budget. Although a business might replace its office with a newer and safer building, older hospital buildings are often kept and expanded. That's a problem because seismic retrofitting is expensive and can cost even more than a new building.

However, the hospital requirement was an unfunded mandate. The state didn't provide the necessary funds to the organizations that run hospitals. Most of them have no way to pay for the retrofits, which can cost as much as their annual budgets. As a result, more than 40 years later, most of the state's hospitals don't meet the standard and at least $50 billion would be needed. The deadlines for meeting the standards are being extended, and there's discussion about modifying the standards. There are additional problems because many hospitals are in poor financial shape, and 20% of Californians, about seven million people, have no health insurance and rely on free care from hospitals. As a doctor told me, "We could treat a lot of people for $50 billion." It costs about $3,000 to insure someone, so $50 billion would cover all the uninsured for about two-and-a-half years. On the other hand, that's about half of what Americans spend gambling every year.

A situation similar to that of hospitals occurs for older concrete buildings, typically with weak first stories, that aren't ductile enough and so could collapse in an earthquake. Although newer buildings are safer, tens of thousands of the older "non-ductile" buildings across California are still used as stores, schools, office buildings, and apartments. Many tenants wouldn't or couldn't pay more for the extra safety, so property owners couldn't pay the high cost of retrofitting these buildings and stay in business. Putting them out of business would reduce the supply of inexpensive housing and office space.

In fact, I lived in one of these buildings when I was in graduate school. I never considered paying more for a safer building and don't recall any of the other seismology students doing so. Instead, I got a top floor apartment and kept my bicycle there in case the building collapsed and crushed the parked cars below.

The issue in these cases is that property owners are told to pay for safety that doesn't directly benefit them. They'd bear the costs, but the benefit of safer buildings is to the whole community. This doesn't work. Instead, the community has to decide how much safety it wants, how much it's willing to pay, and how to pay for it. These are tough and complicated choices, but have to be made.

A further complication is that cost-benefit analysis is just part of the decision-making process for earthquake safety or other issues. Daily life shows that non-economic preferences matter. Many of us get annual physical exams that together cost about $7 billion each year, although studies show that the benefits are less than the costs. We pay for the satisfaction of feeling safer. Conversely, people don't eat enough healthy foods though their benefit is much greater than their real cost. There's an emotional, non-economic cost to eating healthy food rather than junk food.

Like individuals, communities have preferences that influence decisions beyond what's purely logical. In theory, communities could look at different health and safety options and decide which would save the most lives per dollar spent. Researchers who have looked at the decisions society has made find that on average society will spend about $5 million to save a life.

However, people take some high-cost measures and don't take others that cost much less. Much more is spent trying to protect against scary, exotic risks like rare diseases or terrorism than against familiar risks like traffic accidents or common diseases that kill many more people than exotic ones. Only about one-third of U.S. adults get shots to protect against the flu that kills about 35,000 people every year. On the other hand, 50,000 people are paid to screen airline passengers who are very unlikely to be terrorists. People won't give up using cell phones while driving, which would save about 1,000 lives per year, but will stand in long security lines at airports.

As between people, preferences vary between communities. Some will pay higher taxes for good schools, like the Chicago suburbs where I live whose schools are famous (or infamous) from movies like *Ferris Bueller's Day Off.* Others don't think it's worthwhile. Los Angeles, where entertainment is a major industry, spent more than a million dollars on Michael Jackson's funeral while laying off 2,500 teachers. Because of these differences, it's important to address earthquake safety at a local level.

New Madrid Issues

Earthquake safety issues are much more complicated for New Madrid compared to California because it's much harder to decide what the earthquake hazard is and there's little experience with the costs and benefits of earthquake-resistant construction. Deciding what to do needs a lot of careful thought that brings together seismology, engineering, economics, and policy to assess the seismic hazard and choose a level of safety that makes sense.

Building codes are always a balance. Making them too weak lowers costs but allows construction that's less safe and thus higher risk. Making them too stringent imposes unneeded costs on the community and uses resources that could be better used otherwise.

Too-stringent codes can actually produce less safety because—as the adage says—the best can be the enemy of the good. Over time, even if the codes don't change, older buildings will be replaced by newer, safer ones. However, if the codes make new construction more expensive than those planning new buildings think makes sense, older buildings won't be replaced as fast as they would be if the code were less stringent. In addition, builders who would follow codes that seem sensible may evade ones that seem to raise costs without corresponding benefits. This is especially true for governments that can modify or exempt themselves from codes. For example, the bridge at Memphis bringing Interstate 40 across the Mississippi River is designed—sensibly in my view—for a magnitude 7 earthquake rather than the much larger ones assumed in the USGS maps.

All of this is a gamble, but that's life. Balancing how much to spend on earthquake safety compared to schools, hospitals, parks, and other community services is a judgment call for which there is no right or unique answer. Although national codes offer overall insight, local jurisdictions are under no obligation to accept them and will do better by modifying them to balance local hazards and costs.

Unfortunately, these issues weren't thought out when FEMA promulgated a new building code, the 2000 International Building Code. This code, called IBC 2000, would increase the earthquake resistance of new buildings in the New Madrid area to standards like those in California. Surprisingly, the new code was proposed without considering whether the benefits justified the costs.

This puts communities in the area in a situation like a homeowner being pitched a new policy by an insurance agent. The agent says the policy will cost more but offer better coverage. When asked how much it costs, the agent says he doesn't know. When asked how much coverage the policy gives, the agent also doesn't know. Most people would tell the agent to first figure out the costs and benefits and then come back to discuss whether to upgrade their policy.

If pressed FEMA and its supporters claim that every dollar spent in hazard mitigation might save as much as $3—or sometimes even $10—in future repairs. This sounds good but doesn't mean much. First, there are the problems of "might" and "as much as." In a recent weekly Illinois lottery, you might have won as much as $81 million. Still, you need to know the odds to decide if playing is a great deal. Second, the statement is supposed to apply for all mitigation measures anywhere in the U.S. This can't be true because the hazards aren't the same. Hurricane shutters for houses probably make sense in Miami, might make sense in Boston, and don't make sense in Las Vegas. Third, there's no time given for the payoff. An investment that triples your money in 10 years is pretty good, but one that takes 2,500 years isn't. Because buildings have an average life of 50–100 years, most of the money spent on mitigation will never pay off. For all these reasons, it's important to look carefully at the costs and benefits of any specific mitigation policy.

One reason to suspect that the new code might be a bad deal is that the business community has to be pressured to accept it. These people are skilled investors; if spending money to reduce hazards to their buildings made economic sense, they'd probably do it on their own. The fact that they don't want to means that the code probably doesn't make sense for them. However, whether the code makes sense for society as a whole is a more complex but important issue.

Because there wasn't any cost-benefit study for the New Madrid area, Joe Tomasello and I did some simple back-of-the-envelope estimates using the best numbers we could find.

For the cost, we assumed that the new code would increase building costs on average by about 5%. The actual number, which could be a lot more, depends on the type of building and what's done. Because about $2 billion is spent yearly on construction in the Memphis area, that's a $100 million increase per year.

Additional costs, which we didn't include, would be incurred by following FEMA's recommendation to retrofit existing critical buildings like schools, hospitals, fire and police stations, and infrastructure including highways, bridges, utilities, and airport facilities. These costs can be almost as much or more of the cost of a new building. For example, reconstruction of the Memphis Veterans' Administration hospital, including seismic retrofitting, cost $64 million. This involved removing nine floors of the 14-story tower, which made the cost about the same as a new building (fig. 15.6). These funds could have been used for patient care.

This calculation also didn't factor in another important cost of a more stringent code. Although nongovernment buildings wouldn't legally have to

FIGURE 15.6 The Memphis Veterans' Administration hospital during reconstruction and seismic retrofitting. (Joseph Tomasello)

be upgraded, there would be strong pressure to do so. Existing buildings would look unsafe compared to the new code, which would create pressure to tear down or seismically upgrade them. Although this would be good if there were a major danger of a large earthquake soon, it's a needless cost if the danger is small.

For benefit, we started with FEMA's estimate of the annualized earthquake loss in Memphis. That's the number the agency got by dividing how much you expect to lose in a big earthquake by the number of years between them. FEMA got $17 million per year, which is an upper bound because it comes from the USGS maps that predict a hazard on the high end of possible predictions. Thus, even if the code eliminated all future earthquake damage, it would save much less than it costs. In fact, only part of this benefit would happen because even with the new code a large earthquake would cause some damage.

Joe and I got the same answer looking at the benefit another way. This used FEMA's estimate of the annualized earthquake loss ratio, the ratio of annualized earthquake loss to the replacement cost of all buildings in the area. Assuming that this loss is halved by the new code, then over 50 years a building will on average lose 1% of its value, which is less than the approximately 5% cost increase expected from the new code.

In other words, even using FEMA's numbers, which are on the high end because they assume that a major earthquake's on the way, the new code doesn't seem to make economic sense.

FEMA's own numbers also don't support the argument that New Madrid is as dangerous as California. The agency's annualized earthquake loss ratio values for Memphis and St. Louis are about one-fifth to one-tenth of those for San Francisco and Los Angeles. Memphis ranks 32nd in the nation among major cities, and St. Louis ranks 34th (just above Honolulu). These estimates indicate that buildings in the New Madrid region are five to ten times less likely to be damaged during their lives than are buildings in California. That's about the same as estimated in the last chapter by comparing the motion across the faults that cause the earthquakes.

From these estimates, it looks as if the proposed code is likely to cost more than the value of its benefits. Thus, there are probably better ways for communities in the area to use their resources. Building to California standards in the Midwest looks like an expensive cure for the wrong disease. It's like chemotherapy for a cold. Instead of rushing to adopt the new code, it looks more sensible to adopt a less stringent criterion now and conduct detailed studies to decide whether going further makes sense.

At present, states in the area are deciding what to do. The most extensive discussions are happening in Tennessee. Here the state adopted the new building code, which raised the required level of seismic resistance. For example, it requires buildings in Memphis to withstand shaking four times more intense than required by the code the city adopted in 1994. As a result, Memphis and surrounding Shelby County opted for a less stringent version of the new code. This alternative is based on the hazard predicted by the once-in-500-years map, not the once-in-2,500-years map.

I think an approach like this makes sense because large earthquakes in the area are rare enough that typical buildings are very unlikely to be damaged during their lives. The USGS 500-year map still gives a large margin of safety because of the assumptions made in it, notably that major earthquakes are on the way. Because communities face many problems other than earthquakes, I think the resources needed to meet the more stringent version would do communities more good if used otherwise. Over the years, as scientists learn more about the earthquake hazard and new construction methods become available, building codes can be adjusted accordingly.

Chapter 16

What to Do?

NASA owes it to the citizens from whom it asks support to be frank, honest, and informative, so these citizens can make the wisest decisions for the use of their limited resources.

—Physicist Richard Feynman (1918–1988), in a report after the loss of the space shuttle Challenger

---

As we've seen, the New Madrid earthquake issue has become a perfect mess. Some organizations, notably the federal government, have convinced themselves that billions of dollars should be spent to defend against large earthquakes that are on the way. However, the more we learn about the science, the less sense this makes. Now what?

As an old saying goes, "when you're in a hole, stop digging." When something isn't working, stop, think, and come up with a better approach. It's not easy, but it's the only way out. In that spirit, this last chapter gives some ideas that I think would help make more sensible policies.

Tone Down the Hype

The first thing to do is talk calmly about the earthquake issue. There's plenty of public interest in the possibility of large earthquakes in the Midwest, especially around the 200th anniversary of the 1811–1812 earthquakes. This bicentennial gives the public a chance to ask thoughtful questions and expect reasoned answers rather than hype. It's an opportunity to discuss what's known and unknown about the earthquakes and their possible hazards as well as the costs and benefits of different mitigation policies.

A lot can be done in this direction. I had fun working with Chicago's Field Museum of Natural History on an exhibit about natural disasters. Because it was partly for a midwestern audience, we included a section about New Madrid earthquakes. It explained what happened in 1811–1812, how GPS data showed that such earthquakes probably wouldn't happen soon, and raised the question of how much of society's resources should be used to prepare for such earthquakes. The exhibit seemed to do well getting visitors thinking about these issues.

From this and other experiences, I've learned that much of the public welcomes intelligent discussions rather than terrifying predictions of impending disaster. People enjoy applying the healthy skepticism that they use in daily life. Once they start asking tough questions, it's fun to dissect disaster predictions to see what's swept under the rug.

My favorite is a recent report for FEMA from engineers at the Mid-America Earthquake Center at the University of Illinois. This center, incidentally, is different from the school's fine geology department. The report lays out the results of a computer model predicting what would happen if an 1811–1812-style earthquake with a magnitude of 7.7 occurs. It goes on for 936 pages, listing types of buildings damaged, injuries, tons of rubble, and deaths. For example, in Arkansas 37,244 people are looking for shelter, 50,159 buildings are destroyed, and 574 deaths occur. These numbers are a perfect example of the difference between precision, the detailed but often-meaningless numbers coming from a calculation, and accuracy, which is how close the numbers might be to reality. The computer model is amazingly precise, but it's unlikely to be accurate. The problem is that the computer model is based on the very uncertain hazard maps that assume that a major earthquake will happen soon. The report doesn't even mention the GPS data that don't show any large earthquakes on the way.

The report is full of curious numbers. It says "the chance of a magnitude 6 or 7 earthquake in the next 50 years is roughly 90%." Actually, there's an earthquake with a magnitude of 6 or greater in the area about once every 175 years. So in 50 years, that is about a 30% probability. Thus, the report starts with a probability that's about three times too high. Next, as we saw in Chapter 8, big earthquakes are much rarer than smaller ones. Even without the GPS data we'd expect magnitude 7.7 earthquakes to be about 40 times less common, and the GPS data show that one is even less likely. Thus, the report uses a too-high estimate of the probability of a magnitude 6 earthquake to motivate its claims of what would happen in a much rarer and almost 400 times bigger magnitude 7.7 one.

This pitch is like the way state lotteries try to convince you that you have a good chance of winning. They quote the chance of winning, which sounds good. However, that number combines the chances of matching 3 of 6 numbers (1 in 67), 4 of 6 numbers (1 in 311), and 5 of 6 (1 in 73,000) with the chance of matching all 6 (1 in 20 million). The first pays $3 on a $1 bet, but you need to match all 6 to get millions of dollars. There's no harm in planning which airplane, sports car, and yacht to buy if you win, as long as you remember the "if."

Reports of impending doom will keep coming and get media play because people enjoy thinking about disasters. This can be useful: My classes like videos of real earthquake, tsunami, and volcano damage. However, I also discuss how likely the videos are. Some of the most exciting fictional ones are impossible, like the lava destroying downtown Los Angeles in the film *Volcano*. As a review on the Web site *badmovies.org* explains, "If you believe this film bears any resemblance to reality, it's probably because you live in LA."

"Disaster chic" is fun and harmless unless people believe the fears. Unfortunately, this sometimes happens. As Simon Briscoe and Hugh Aldersey-Williams explain in their book *Panicology,* "there are serious emotional, social and economic costs to panic . . . We'd be a lot happier if we insisted that people prove their case before making dire pronouncements. We shouldn't be wasting time worrying about a lot of stupid things."

The best response to disaster predictions is simple common sense. Financial planners advise clients to avoid scams by remembering that if something sounds too good to be true, it probably is. Similarly, when a predicted disaster sounds too terrible to be true, it probably is.

In dealing with disaster predictions, the media can help or hurt the public. People both enjoy disaster stories and have healthy skepticism. Thus, although it's fun to run scary stories, it would be more useful to ask hard questions that reinforce the public's sensible instincts. As we saw in Chapter 2, thoughtful reporting would have avoided the debacle of the Iben Browning earthquake prediction. One simple solution would be for New Madrid earthquake disaster reports to have a warning label like those on cigarette packs. One that would help people decide how much to worry about the predicted catastrophe could be

WARNING: Although large earthquakes have happened here in the past, geophysical data show no sign that a large earthquake is coming soon.

Use What We Know

According to President Ronald Reagan, "the nine most terrifying words in the English language are: 'I'm from the government and I'm here to help.'" This statement reflects the fact that Americans feel that many government policies don't make sense. The federal government is an enormous presence in our lives. It spends $3.5 trillion a year, which is about 25% of the gross domestic product. We spend 10 billion hours a year filling out its forms. Our views on what all this gets us are mixed. Many people think that the government does some things well, like running national parks. However, there's a strong sense that many government activities involve expensive and poorly thought out programs or regulations to deal with problems that are either exaggerated or don't exist.

These views come out in national polls. When asked, "How much of the time do you think you can trust the government in Washington to do what is right?" 2% of those surveyed say "always," 21% say "most of the time," 68% say "only some of the time," and 7% say "never." When asked "Would you say the government is pretty much run by a few big interests looking out for themselves or that it is run for the benefit of all the people?" 64% say "a few big interests," and 28% say "all the people." Hence, 55% of people think of "the" government, whereas only 42% think of "our" government.

This skepticism crosses political lines. A joint study done by the Democratic-leaning Brookings Institution and the Republican-leaning American Enterprise Institute concluded that although the cost of federal environmental, health, and safety regulations was hard to measure, it was about as much as everything in the federal budget excluding defense and mandatory spending like Social Security. In their view:

> The benefits of these regulations are even less certain . . . Research suggests that more than half of the federal government's regulations would fail a strict

> benefit-cost test using the government's own numbers. Moreover, there is ample research suggesting that regulation could be significantly improved, so we could save more lives with fewer resources. One study found that a reallocation of mandated expenditures toward those regulations with the highest payoff to society could save as many as 60,000 more lives a year at no additional cost.

When government policies don't make sense, people have two choices. Usually, we're fatalistic. We accept pointless policies because it seems hopeless to try to change them. We fill out useless and incomprehensible forms, stand in line to renew driver's licenses that could be done online, and take off shoes in airports. However, sometimes we take an idealistic view and try to change government policies.

As you might guess from the fact that I wrote this book, I tend toward the second view. That's mostly because I've been active in environmental causes in a small way for years and have done the usual letter writing, phone calling, petition signing, etc. This might have helped on the national and state levels: Some things I thought were good got done, and some bad things were stopped. It definitely worked once on the local level. I still enjoy walking by a small woodland that was going to be cleared for housing. My neighbors and I saved the site after years of petitioning, attending village board meetings, letter writing, and the like.

Based on these experiences, I believe that change starts with the public. If Midwest communities take a calm view of earthquake hazard issues, they might convince the government to follow. Changing the hazard policy won't be easy, will take time, and might not happen. Still, I think it's worth trying.

The view of the Midwest earthquake hazard in the geological community is changing rapidly because science is set up for change. That's because science has a single goal: to understand how nature works. Scientists know that some ideas turn out wrong, and some won't lead anywhere, but some will help us understand the world better. Over time we both expect new data to change our ideas for the better and accept that these won't be the final answer.

Governments are the opposite; they're set up to make sure things stay the same or change slowly. That's why they have complicated laws, regulations, and bureaucracy. The huge effort required for even trivial changes, like those of the Nixon, Carter, Reagan, and Clinton administrations to abolish the board of tea tasters established in 1897, pretty much makes government programs immortal.

Thus, there's a big cultural difference between science and government. In politics, a lot of effort goes into debating what the Founding Fathers meant when they wrote the Constitution in 1787 and changing it as little as possible. In contrast, scientists don't worry about what Newton thought in the late 1600s; we use what he got right but have discarded what he got wrong.

The issue of earthquakes in the central U.S. is an example of the opposite ways science and government work. Over the past 20 years, the government has been ramping up efforts to deal with what it has convinced itself is a major earthquake hazard. Over the same time, using new data, science has been going the other way. The more we look, the smaller the hazard seems. Thus, the government wants to invest lots of resources to "solve" a problem that's much smaller than they think and could be very minor. I don't know why the government is acting so aggressively relative to Midwest earthquakes, where GPS data show that nothing's happening, while largely ignoring more serious problems.

The government's New Madrid policy made some sense 20 years ago because all science could say was that there had been major earthquakes in the past, so they might happen again. Now that GPS data give a very different picture, the new science could be used to make more sensible policy. Despite the government's reluctance to change, it would be silly not to.

Discuss Uncertainties Fully

Making earthquake policy for the New Madrid area is an example of the challenge society faces when deciding what to do about a possible future problem. Because this happens often, there's been a lot of thinking about how to decide. The major lesson is that the decision can't be made for society by "experts" because there really aren't any. People who study problems can tell what they know and what they think, but no one is sure what will happen.

It's now clear that the best way for a community to make sensible policy is to start with open discussion of what's known and what isn't, including the uncertainties. A nice recommendation comes from *Prediction: Science, Decision Making, and the Future of Nature* by Daniel Sarewitz et al.

> Above all, users of predictions, along with other stakeholders in the prediction process, must question predictions. For this questioning to be effective, predictions must be as transparent as possible to the user. In particular, assumptions, model limitations, and weaknesses in input data should be forthrightly discussed. Institutional motives must be questioned and revealed . . . The prediction process must be open to external scrutiny.
>
> Openness is important for many reasons but perhaps the most interesting and least obvious is that the technical products of predictions are likely to be "better"—both more robust scientifically and more effectively integrated into the democratic process—when predictive research is subjected to the tough love of democratic discourse . . .
>
> Uncertainties must be clearly understood and articulated by scientists, so users understand their implications. If scientists do not understand the uncertainties—which is often the case—they must say so. Failure to understand and articulate uncertainties contributes to poor decisions that undermine relations among scientists and policy makers.

Although this hasn't happened in New Madrid earthquake policy making, the atmospheric sciences show what should be done. Meteorologists are much more candid about hazards due to weather, which in the U.S. causes about 500 deaths per year compared to about 20 per year due to earthquakes. One key is comparing predictions by different groups using different assumptions. For example, on February 2, 2000, the *Chicago Tribune* weather page stated:

> Weather offices from downstate Illinois to Ohio advised residents of the potential for accumulating snow beginning next Friday. But forecasters were careful to communicate a degree of uncertainty on the storm's precise track, which is crucial in determining how much and where the heaviest snow will fall. Variations in predicted storm tracks occur in part because different computer models can infer upper winds and temperatures over the relatively data-sparse open Pacific differently. Studies suggest that examining a group of projected paths and storm intensities—rather than just one—helps reduce forecast errors.

The newspaper's graphics compared four models' predicted storm tracks across the Midwest and seven precipitation estimates for Chicago. This nicely explained the models' uncertainties, their limitations due to sparse data, and the varying predictions.

Similarly, on a monthly time scale, the National Oceanic and Atmospheric Administration's Climate Prediction Center compares various university and government predictions seeking to describe events that may have billions of dollars in economic impact. It cautions "potential users of this predictive information that they can expect only modest skill."

Moreover, analyses of the possible effects of global warming compare different models developed by various groups. The United Nations Environment Program notes that "there remains substantial uncertainty in the exact magnitude of projected globally averaged temperature rise caused by human activity, due to shortcomings in the climate models . . . Furthermore, scientists have little confidence in the climate changes they project at a local level. Other uncertainties, not arising from specific limitations in the climate models, also restrict the ability to predict precisely how the climate will change in the future."

Discussions of earthquake hazards should be as clear about what is and isn't known. Comparing alternative hazard models would illustrate the uncertainties. The discussions should also explain that the real uncertainties are almost certainly bigger because of effects that weren't considered and factors that were treated as well known but aren't. Although it can be embarrassing to admit how little is known, candor would help communities make much better decisions.

Keep Thumbs off the Scale

When planning in a risky situation, people often protect themselves from future blame by projecting unlikely worst-case scenarios. This tendency to bias estimates is called "conservatism" by its supporters and "CYAing"—covering your ass—by its detractors. Although CYAing sounds like a good idea, history shows that it can cause problems.

Often in military planning, one side convinces itself that the other is "10 feet tall" and so makes bad strategy choices. In the Civil War, Union General George McClellan often convinced himself that the Southern forces were much larger than they actually were and refused to act when victory was possible. His passivity led President Lincoln to say, "If General McClellan does not want to use the army, I would like to borrow it for a time." During the Cold War, there was the idea of a "missile gap," in which the Soviet Union's nuclear force was supposed to be more powerful than the U.S.'s. Although President Eisenhower pointed out correctly that the U.S. force was far superior, fear took root and led the Kennedy administration to an expensive buildup that escalated the arms race. Today, the Department of Homeland Security, FEMA's parent agency, assumes that terrorists have almost magical powers. In the department's view, "terrorists can strike at any place at any time with virtually any weapons."

In natural hazard analysis, extreme assumptions can be made at many stages. These amplify each other to seriously distort the result. As we've seen, this happened for the New Madrid earthquake hazard. The hazard maps use high values for the probability that large earthquakes will happen, high values for their magnitudes, and high values for the shaking that will result. These together gave the questionable prediction that New Madrid is more dangerous than California. Such biased estimates distort policy decisions by favoring earthquake safety over other uses of resources that could do more good.

Make Policy Openly

Discussing uncertainties openly helps communities consider alternative strategies. That's crucial because there are no right answers. State and local governments have to weigh many factors and decide what to do. I think these tough decisions should be made at the state and local level, not forced down from Washington. That's because communities, not the federal government, will bear most of the costs. It should be their decision how to balance costs and benefits and make the difficult trade-offs between them.

A good example is the process used in Tennessee to decide whether to strengthen the 1999 building codes in the Memphis area to the newer IBC 2000 code.

There was considerable discussion in the media. The Memphis *Commercial Appeal* explored the issue in news coverage and editorially. It noted, "Don't expect state and local governments to follow Washington's lead at the

Memphis Veterans Medical Center, which is undergoing a $100 million retrofit to protect the building against the potential threat of an earthquake. Unlike the federal government, local governments are required to balance their budgets, and expenditures of that size would be hard to justify."

The relevant state agency then held public hearings in response to a request by the mayors of Memphis and surrounding Shelby County. Different stakeholders got the opportunity to give their views. The USGS, FEMA, and supporters of the stronger code gave their traditional arguments. However, the broader perspectives from the community that were ignored in developing the code were also thoughtfully discussed.

Memphis Mayor Willie Herenton pointed out the need for careful thought: "With the potential damage this code can cause to our tax base and economy we should have complete data and ample opportunity to make the decision. Our community has financial dilemmas that necessitate making wise choices. We cannot fund all of the things that are needed. Is it appropriate for us to spend the extra public funds for the enhanced level of seismic protection above and beyond our current codes? Would we be better off paying higher salaries for our teachers and building classrooms? This is the kind of public policy debate that should occur before new codes are warranted."

Bill Revell, the mayor of nearby Dyersburg, a major manufacturing community, expressed concerns about economic development. He explained that a 10–15% increase in building cost would hurt the town's ability to attract industry because "they look at everything. They look at tax incentives. They look at the labor force. They look at the cost of construction. And this would be very detrimental to us."

Architect Kirk Bobo explained, "I do believe that the added cost for seismic design is material. We do work all over the country, California included. And we have probably more than average knowledge of cost implications of seismic design. And I believe further that the impact of non-construction, direct construction-related economic impact, and the cost implications of that, is something that warrants a tremendous amount of analysis. I don't believe that has been done, and I think it's critical that it be done before making an informed decision."

Architect Lee Askew echoed these thoughts: "There is a premium to constructing buildings with this advanced level of seismic design. It will affect structure, certainly. It will heavily affect some of the other systems including mechanical, electrical, plumbing, etc. . . . As I assist these large national clients, I see how competitive it can be from one community to the next, one state to the next. If we are doing something to hobble ourselves, then we will certainly be disadvantaged."

Developer Nick Clark pointed out that the proposed costs for reducing earthquake losses were much greater than used to fight West Nile virus, which had caused four deaths in that past year in the county, despite the much lower earthquake risk. More generally, he argued, "We have needs in our community that are much greater in terms of the investment in human life. We have an educational system that is broke, that needs to be fixed . . . This is about seismic issues in earthquakes. But there's an earthquake that can happen in terms of the construction of our society if we cannot take care of our citizens and train them to be healthy and productive individuals. There needs to be a connection between how we handle these resources in terms of the investment in citizens of Shelby County."

I think the resulting decision, to modify the proposed IBC 2000 code, was wise. More important, regardless of the outcome, this discussion is a good model for other communities.

Make Policy Thoughtfully

Making a good decision usually involves taking the time to get it right. As the old saying goes, "Do you want it right, or do you want it now?"

In this spirit, I think the best way to make sensible policy for Midwest earthquakes is take a deep breath, calm down, and think carefully about what to do. There's no reason to believe that raising construction standards to the California level is worth the billions of dollars it would cost. Until this issue is settled, a rushed decision to adopt the new codes would likely be wrong. Usually, haste makes waste.

A better approach would be for the states in the area to commission an objective outside study to assess costs and benefits of different policies. That's complicated because there are different estimates of the hazard and different possible strategies to mitigate it. A serious study involves seismology, engineering, and economics. However, some consulting firms have the expertise to do a good job. It wouldn't be cheap, but spending a few million dollars would pay because the construction costs under consideration over many years are billions of dollars.

The states would have to carefully choose the questions they want explored. There's a wide range of proposed hazard levels, which would require different strategies.

The high-end hazard model is the traditional USGS/FEMA one. In this, the 1811–1812 earthquakes will recur in the next few hundred years, so the highest hazard is near the faults that broke then. The estimates Joe Tomasello and I did imply that even for this scenario, the California-level code is unlikely to be worthwhile because it calls for building structures with a typical 50–100 year life to resist the maximum shaking predicted on average once in 2,500 years. Thus, the study should also evaluate building to the less expensive levels required to withstand the maximum shaking predicted on average once in 500 years. It would also be worth considering other strategies.

The lower hazard model comes from the GPS data that show no sign of a major 1811–1812-type earthquake on the way. There's still some hazard from smaller earthquakes, up to a magnitude of about 6.7 Most of these seem to be aftershocks, so they're more common near the faults that broke in 1811–1812, but some occur across the whole region. Because these aren't that big, they're unlikely to do serious damage unless by bad luck they happen close to a heavily populated area. Although a building code could reduce this damage, it's worth trying to figure out a code level that wouldn't cost much more than it accomplishes. As we've seen, the issue is that any building is very unlikely to be shaken seriously during its life.

It would also be worth studying other options. Even if a stringent code doesn't make sense for ordinary new construction, some things might be worth doing. One is a high standard for critical facilities. Jay Crandell, an engineer who studies earthquake damage, suggests that focusing mitigation efforts on existing unreinforced buildings would reduce the risk to life at modest cost.

The study should consider different hazard models and perhaps a blend of them like a logic tree. For each, it should explore a menu of options with the estimated cost and benefit. Naturally, these would have large uncertainties. Still, this analysis would be a huge step forward from what has been done so far, which is basically nothing. It would also be a start for future studies because it could be upgraded as scientists learn more.

As well as commissioning this study, the states should develop expertise in earthquake hazard modeling so they can best use the results. Having scientists who understand the scientific, economic, and policy issues focus on a state's interests is crucial. The fact that the California state geological survey has this expertise is part of the reason why California is always "at the table" when the federal government proposes earthquake policies. Thus, the new codes didn't impose the very expensive, once-in-2,500-years criterion on California that they required in the Midwest.

In contrast, most midwestern states' geological surveys don't have earthquake hazard modeling expertise because damaging earthquakes are rare. This expertise is valuable: Zhenming Wang and his colleagues at Kentucky's survey have been evaluating the effects of proposed FEMA and USGS policies and proposing changes to reflect the state's interests.

Expertise within state government would also give officials and community groups additional advice. Generally, they have had to rely on outside scientists. In several situations I've been advising business groups while other scientists have been funded by the USGS. There's no harm in this, but I'm sure community groups would appreciate having scientists involved whose primary responsibility was to the state and were familiar with the state's needs and interests. The cost of hiring good people would more than pay for itself in the benefits to the state's economy.

Whatever is decided now shouldn't be viewed as final. Our understanding of the science will evolve with time, so we'll be able to make better hazard models. For example, if we see significant strain starting to build up, it might make sense to take additional mitigation steps. Technological changes will make mitigation cheaper, so measures that don't make sense now might make sense eventually. Communities will become more earthquake-resistant even without upgrading codes as older buildings are replaced by safer ones. Thus, over time communities will want to reassess and perhaps change their earthquake hazard policies.

Improve Disaster Planning

The damage to New Orleans and surroundings by Hurricane Katrina showed the weaknesses in our systems for dealing with natural disasters. This a tough challenge that has four major components: assessing the hazard from possible future disasters, developing strategies to mitigate their effects, responding to the immediate effects of a disaster, and rebuilding after the disaster.

The four components involve different bodies of knowledge and skill. Hazard assessment is the "cleanest" because it's a scientific issue. Hazard mitigation is the "messiest" because it involves science, engineering, economics, policy, and politics.

For Katrina, hazard assessment worked well. It predicted both the likelihood of the storm and the resulting flooding. However, the other three components failed. Mitigation measures proved inadequate, emergency response was poor, and rebuilding has been very slow. Thus, there's been a lot of interest in looking back and using the lessons learned to do better.

In this book, we're going the other way, looking into the future. We've discussed how likely a major New Madrid earthquake is within a few hundred years and what to do about this possibility. These two components of disaster preparation are the ones that involve science.

As we've seen, neither has been well handled by FEMA and associated agencies. For what the public spends on them, we should expect much more. The question is how to do better. From my view as a scientist, it's easy to identify some of the possible changes that would improve the process. Whether they can be done politically is a tough question because government bureaucracies are difficult to reform. Even so, it's interesting to consider possible approaches.

The New Madrid situation shows several general problems. First, the hazard assessment ignores new science. Second, it overemphasizes one hazard, earthquakes, compared to others in the same area, like floods and tornados, which are more serious. Third, it doesn't consider the costs and benefits of the proposed mitigation strategy to communities in the area. Fourth, it was imposed by Washington rather than developed jointly with those communities.

A better approach would be for Congress to require FEMA or another agency to develop the expertise to make an integrated natural hazard policy for earthquakes, floods, storms, wildfire, drought, etc. This would involve a national assessment of the different hazards and a strategy to mitigate them. Both the assessment and mitigation plans would be for 10 years and developed under strict criteria. The law would require the assessment to be based on up-to-date science from academia and government. Similarly, the mitigation plan would be required to be developed jointly with state governments, community groups, and the private sector, and to consider the costs and benefits of alternative strategies. The assessment and mitigation plans would have to consider results and policies from other countries, which are often ahead of those in the U.S.

Review committees whose members would not be federal employees would approve both plans. The assessment review committee would be scientists with expertise in the different natural hazards, and the mitigation review committee would include engineers, economists, and representatives of state governments, the private sector, and community groups.

This approach draws on two ideas from defense planning, which has some similarities to hazard planning. One is Congress's requirement that the Department of Defense conduct a major study every four years called the Quadrennial Defense Review. The QDR assesses likely requirements in the coming years and how U.S. strategy, programs, and resources should be allocated to meet them within the forecast budget plan. The review recognizes that strategies should change as situations change and consider budgetary constraints. Although no planning is perfect, the QDR process helped the military shift focus after the Cold War.

Defense planning also shows that the best scientific advice comes from people at "arm's length." This was critical to the success of the U.S. military research effort in World War II. Although its goals were military, the Office of Scientific Research and Development was under civilian control, and its scientists and engineers stayed civilians. These choices are credited with developing crucial weapons like radar much faster than would have happened otherwise.

This approach would improve hazard planning by making better use of expertise in universities, state and local agencies, and the private sector. It would also help government science. At present, agencies like the USGS both do science and are involved with political and policy documents like hazard maps and building codes. This puts their scientists under implied—perhaps even unconscious—pressure to come up with results supporting the agency's policy, which discourages new ideas and slows progress. It would make more sense for agencies like the USGS to focus on science and for other organizations to combine their results with those from the broader scientific community. Separating science from policy would be better for both sides. Some senior USGS scientists have told me they're uncomfortable with their agency making hazard maps and would prefer to focus on the science that goes into them.

A consistent assessment and mitigation plan for different hazards would have many advantages. Because today each hazard is treated arbitrarily and differently, there's no consistency. FEMA wants buildings built for the maximum wind expected once every 50 years, the typical life of a building, which there's a 2% chance of having in any year. However, the agency tells communities to plan for what's called the 100-year flood, which is a higher standard. This is the maximum flooding expected on average once every 100 years, or that there's a 1% chance of having in any year. FEMA wants even higher standards for earthquakes. California should plan for the maximum earthquake shaking expected on average once in 500 years, and midwestern states should plan for the maximum shaking expected on average once in 2,500 years. This pattern is the opposite of what we'd expect because wind and flooding—often due to the same storm—cause much more damage than earthquakes.

None of these time periods comes from careful analysis, and it's not clear which, if any, should be different. It might better to plan for both 500-year floods and earthquakes. We've seen that using 2,500 years is likely to overprepare for earthquakes. Conversely, it seems that in many areas planning only for the 100-year flood gives too low a level of protection, so it would be wise to prepare for larger floods.

There's an interesting difference between floods and earthquakes. It looks like the New Madrid fault system is shutting down for a while, so the historical record overpredicts the danger. However, the historical record of flooding probably underpredicts the danger. Major floods might be getting more common because the more people build on flood plains, the more water ends up in rivers to flood downstream. It's also possible that as the climate warms, coastal flooding will be a larger problem if major storms like hurricanes become bigger or more common.

A reality check would be for Congress to require that the mitigation plan be advice, not a requirement, for state and local governments. Part of the reason many current rules aren't carefully thought out is that they don't have to be. That's because FEMA applies financial pressure to state and local governments to adopt its policies. Eliminating the ability to twist arms will force federal policies to be well conceived enough that they're adopted voluntarily. This approach works in medicine: Recommendations from groups like the American Academy of Pediatrics don't have legal force but are generally followed because they're usually sensible approaches coming from a lot of careful thought.

An integrated hazard policy would also help address the problem of natural disaster insurance. After a disaster, insurance compensates property owners for damage and thus provides funds for rebuilding. However, major disasters could overwhelm insurance companies' ability to pay claims. Insurance companies work by spreading out risk. For example, only a small fraction of homes in an area will burn down in any year, and that fraction is about the same every year. In contrast, a natural disaster can cause a huge number of claims at once in one area. Thus, insurance companies have to build up huge financial reserves to prepare for rare events with impacts that are hard to predict. This situation causes problems in hurricane-prone areas. Policyholders have trouble collecting after hurricanes, and some insurance companies are refusing to sell more policies in areas like Florida.

A proposed solution is for the federal government to provide disaster insurance. For example, the New Zealand government insures property owners against earthquakes, landslides, volcanic eruptions, tsunamis, storms, flood, or fire caused by any of these. The problem is that the U.S. is so large that different places have big differences in hazards. There's concern that taxpayers in areas with few disasters would subsidize residents of disaster-prone areas. People in the Midwest would be paying for earthquake damage in California and hurricane damage in Florida. However, a scientifically credible and consistent national hazard assessment would allow disaster insurance premiums that fairly reflect the different hazard levels in different places.

Continue Learning

In the past 20 years, our view of earthquakes in the New Madrid region has changed dramatically. New data have shifted our view from one in which large earthquakes keep happening more or less regularly on the same faults to one in which earthquakes move between many faults that turn on and off. This new science is also changing ideas about the earthquake hazard.

Looking to the future, I think the new "episodic, migrating, and clustered" view puts us on the right path. Still, there is a long way to go. It's important to understand what causes the earthquakes and how they move around between faults. There's also the possibility that everything we're doing today is still missing some important things.

A quotation from former Secretary of Defense Donald Rumsfeld describes this situation. He explained that in making military policy, "There are known knowns. These are things we know that we know. There are known unknowns. That is to say, there are things that we know we don't know. But there are also unknown unknowns. There are things we don't know we don't know."

Thus, we should both use what we know and keep learning more. More science is the key to progress. A lot of research is going on, some of which is bound to yield valuable information. This includes studies that give new data and others that use both existing and new data to try to understand the underlying processes.

Studies of the earthquakes in the region will continue. Although most earthquakes are small, seismic networks can locate them and thus improve maps of earthquake activity. Seismology can also study the geometry of faulting in the rare larger ones, perhaps magnitude 5 or above. As seismometers and methods of analyzing seismograms get better, they yield information from progressively smaller earthquakes. More will be learned about earthquakes on or near the faults that broke in the big 1811–1812 earthquakes and about other faults in the broader area.

GPS data will continue to show what is or isn't happening. Although GPS data in the New Madrid area show no or very slow motion during the past 20 years, that could change. It's also important to see if motion occurs somewhere else in the region. It would be nice to do this with continuous GPS at sites with high-quality monuments that give the best measurements. If that's not practical over the whole area with the available funding, a lot can be done cheaply. One way is to do GPS surveys in which sites are occupied for a few days every few years. Another way is to use continuous GPS sites that the National Geodetic Survey runs. Although the monuments aren't great, the data are free and pretty good. They're what we used to show the beautiful pattern of motion today from the melting of ice age glaciers.

Geological studies of faults and paleoearthquakes on them will continue to give a view into the past that we can't get any other way. It's important to know how faults have moved over time.

Hopefully, we'll also learn more about what causes the earthquakes. Geological studies will give more insight into the complicated geological jigsaw puzzle making up the central U.S. and thus how the different parts have worked over time. There should also be valuable insight from a new seismic network called USArray. USArray has an array of 400 high-quality, portable seismometers called Bigfoot that is moving across the nation in a series of major steps. Additional seismometers will also operate to look at specific regions. They'll "see" deeper than previously possible and give a better view of what's underneath the region. This will give a better idea about what causes the earthquakes and what the differences are between geological structures that have large earthquakes and those that don't. For example, my colleague Suzan Van der Lee is leading a project in which we'll be working with Washington University and the University of Minnesota to study the Mid-continent rift.

The new data will help researchers continue developing and testing models for what causes the earthquakes and how they move around between faults. The resulting new models will, in turn, improve our ability to assess hazards.

In addition to continuing the kinds of research going on today, future researchers might have entirely new concepts and types of data. What's unimagined or impossible today may become normal or possible. Although this sounds far-fetched, it often happens. Fifty years ago no one had thought of plate tectonics. Thirty years ago no one anticipated that GPS would be able to measure ground motions slower than a millimeter per year and thus see if strain was or wasn't building for future earthquakes. Thus when someone writes a book like this 50 years from now, the picture might differ in many important ways.

When my coworkers and I started a GPS study in 1991, we knew it was a long-term project. I told the graduate students that we were laying groundwork for their graduate students' graduate students. Although we and other researchers have learned a lot since, there's still much to learn. Future geologists studying these problems will probably know much more and be amused at what my generation missed. Even so, I expect that they'll face important, challenging, and unsolved questions. I hope this book helps motivate some talented young people into careers in the earth sciences. The best outcome is if one of them makes a major advance in understanding mid-continent earthquakes.

FURTHER READING AND SOURCES

Further information on the scientific topics discussed here is more easily accessible from textbooks than from research papers written tersely by scientists for scientists familiar with the topics under discussion. I've listed several textbooks for specific chapters. Research papers mentioned, including ones from which a figure is used, are listed in the References section by author(s).

I have drawn heavily from books describing the New Madrid and San Francisco earthquakes and on the history of earthquake research and policy. The Internet also contains a lot of information that's convenient but sometimes unreliable. Online information on technical topics, such as high-precision GPS or earthquake-resistant construction, is often excellent. In addition, many primary sources, including newspapers and historical accounts of the New Madrid and San Francisco earthquakes, are available online. Thus, it's often easier and faster to find quotations online rather than to look in books. Similarly, many of the documents listed here are online.

However, because Web sites are easily created and copied from each other without external review, many contain information that's wrong or out of date. For example, a Google search found more than 32,000 references, including the online encyclopedia Wikipedia, to the incorrect legend that New Madrid earthquakes rang church bells in Boston. Searching for the outdated magnitude 8 value for these earthquakes yields about 35,000 results, again including Wikipedia. Problems such as these are why some universities don't let students use Wikipedia as a reference in term papers or theses and students are taught in general to approach online sources cautiously (see *http://www.lib.berkeley.edu/TeachingLib/Guides/Internet/Evaluate.html*). For example, a quotation in an article on a newspaper's site is likely accurate, whereas information on a site discussing the end of the world in 2012 is unlikely to be.

Chapter 1

The epigram is from Feynman (2000). An excellent source about applying high-precision GPS for earthquake studies is the UNAVCO Web site, *http://www.unavco.org*, and its links. Two of the best introductions to the GPS method are *http://www.colorado.edu/geography/gcraft/notes/gps/gps_f.html* and *http://www.trimble.com/gps/index.shtml*.

Chapter 2

The city of New Madrid's Web site (*http://www.new-madrid.mo.us*) describes the community. Newspaper stories generally accessible online, such as Hernon and Allen (1990), describe the zany events following Iben Browning's prediction. Quotations are from the July 25, 1990, *St. Louis Post-Dispatch*, Allen (1990), Byrne (1990), Feder (1990), and Robbins (1990). Kerr's (1990, 1991) columns in *Science* magazine give thoughtful overviews. The Richter quotation is from Hough's (2007) biography.

Chapter 3

The Jackson epigram and the Richter quotations are from Fischman (1992) and Hough (2007). Table 1 is from Stein and Wysession (2003). Data in figure 3.1 are from *http://earthquake.usgs.gov/regional/states/us_deaths.php*. Quotations about swine flu are from Brown (2002). The *Newsweek* Y2K cover story ran on June 2, 1997. Senator Moynihan's letter is in the Congressional Record (August 11, 1996) and available online. Many of the dire warnings and plans can still be found online by searching "FEMA Y2K" and the like. Nimoy's Y2K video is at *http://www.youtube.com/watch?v=cL5Yu3Ub9nA*. Dutton (2010) summarizes how the world as we know it didn't end. O'Connor (2008) reviews the history of the human chromosome number. A history of speed of light measurements is in Henrion and Fischhoff (1986). A video of the gorilla walking through the basketball game is at *http://viscog.beckman.illinois.edu/media/ig.html*.

Chapter 4

Mitchell's quotation in the epigram is from his 1998 memorial to Nuttli. Stewart's role is described by Allen (1990). The Alaskan earthquake and other USGS photos in the book are from *http://libraryphoto.cr.usgs.gov*. Accounts of the USGS program are from Geschwind (2001) and Hough (2007). Kerr (1984, 1985, 1986, 1992, 1993, 2004) and Finn (1992) describe the Parkfield prediction's rise and fall. The *Economist* quotation is from August 1, 1987. The USGS report following the unsuccessful prediction is Hager et al. (1994). Geller et al. (1997) review the general failure of earthquake predictions. Wilkinson's "catastrophic earthquake" claims are from a January 11, 2000, CUSEC press release and his "when not if" assertion is reported in *http://newmadridawareness.blogspot.com*. Brown's "tweak" quotation appears in many news sources. The aquarium story is in the November 22, 2007, *New York Times*.

Chapter 5

I have drawn primarily on Penick (1981) for the history of New Madrid and accounts of the earthquakes. Many primary sources are available at *http://www.ceri.memphis.edu/compendium* and *http://pasadena.wr.usgs.gov/office/hough/page.nm.html.* Other accounts of the earthquakes include Sieh and LeVay (1998) and Hough and Bilham (2006). Church bells ringing in Boston are described in sources including a USGS "Fact Sheet" (Schweig et al., 1995), and references to Tecumseh's prophecy include Hamilton and Johnston (1991). The most recent magnitude values are discussed by Lovett (2010).

Chapter 6

Winchester (2005) is the most recent of the books describing the San Francisco earthquake. I have drawn on Geschwind's (2001) discussions of the state commission's formation, activities, and post-earthquake policy issues. The extensive historical and scientific online resources include

http://www.exploratorium.edu/faultline/great/1906
http://earthquake.usgs.gov/regional/nca/1906/18april
http://www.sfmuseum.org/1906/06.html
http://www.archives.gov/legislative/features/sf
http://bancroft.berkeley.edu/collections/earthquakeandfire
The commission report (Lawson, 1908) is available online:
http://content.cdlib.org/view?docId=hb1h4n989f&brand=eqf&chunk.id=meta
and the photos are at
http://libraryphoto.cr.usgs.gov

Chapters 7–10

Plate tectonics is discussed in introductory geology texts such as Marshak (2007). Popular books including Bolt (2006), Brumbaugh (2010), Hough (2002), and Sieh and LeVay (1998) discuss earthquakes in general and specific ones. Stein and Wysession (2003) discuss these topics in more mathematical and physical depth. Online resources include *http://www.seismo.unr.edu/ep/nvguide* (Nevada Bureau of Mines and Geology, 2000), *http://www.iris.edu*, *http://www.scec.org/education/*, and *http://earthquake.usgs.gov/learn*. Figure 7.9 is modified from *http://pasadena.wr.usgs.gov/office/hough/east-vs-west.jpg*.

Chapter 11

Geophysical methods for exploring the subsurface are discussed in texts including Kearey and Brooks (2002) and Musset and Kahn (2000), and online at *http://www.mssu.edu/seg-vm*. Van Arsdale (2009) reviews the geological history of the area.

Chapter 12

References for geodetic GPS are listed under Chapter 1 of this section. Kerr (1999) describes the AGU meeting. The New York results are discussed by Snay (1986). The results from resurveying the Stanford network are at *http://seismo.berkeley.edu/annual_report/ar02_03/node34.html*. Calais's comment is reported in the December 15, 2005, *St. Louis Post-Dispatch*.

Chapter 13

Tuttle (2001) summarizes New Madrid paleoseismic data. A number of popular books discuss complex systems, including Mitchell (2009).

Chapter 14

Weinstein and Adam (2008) show how scientists use approximation to explore physical problems. The veiled simplicity quotation is from Hanks and Cornell (1994). USGS earthquake hazard mapmaking, including the "not comfortable" quotation, is described in Frankel et al. (1996). These maps and references are available at *http://earthquake.usgs.gov/hazards*. Stein (2009) gives an introduction to hazard mapping. Earthquake probability models are discussed in Stein and Wysession (2003). Papers debating the New Madrid maps include Stein et al. (2003), Frankel (2004), Stein (2005), Wang et al. (2005), and Searer et al. (2007). Alternative maps are shown in Newman et al. (2001) and Hebden and Stein (2009).

Chapter 15

General explanations of earthquake-resistant construction are given in Gere and Shah (1984) and Levy and Salvadori (1992). More detailed discussions are in Coburn and Spence (2002) and FEMA (2006). Web resources include *http://www.celebratingeqsafety.com* and the video from Banda Aceh at *http://www.youtube.com/watch?v=JPj41JySMLk*. Singer (2009) discusses how government agencies value lives. Retrofit issues in California are discussed by the California Health Care Foundation (2007) for hospitals and Bernstein (2005) for older concrete buildings. Charlier (2003) describes the Memphis VA hospital reconstruction. Earthquake loss estimates for cities are from FEMA (2000).

Chapter 16

The epigram is from Feynman (1988). The FEMA report is Elnashai et al. (2008). National poll results are from CBS (August 2009), Council for Excellence in Government (May 1999), and CBS/*New York Times* (July 2004) and are tabulated in *http://www.pollingreport.com/institut.htm*.

The report on the costs and benefits of federal regulations is Hahn and Litan (1998). The Memphis *Commercial Appeal* editorial is from May 29, 2003, and quotations from the building code hearing on August 27, 2003, are from the transcript of that hearing. The Rumsfeld quotation is from a February 12, 2002, Defense Department briefing.

REFERENCES

Aki, K. Presidential address. *Bulletin of the Seismological Society of America* 70 (1980): 1969–1976.

Aldersey-Williams, H. and S. Briscoe. *Panicology: Two Statisticians Explain What's Worth Worrying About (and What's Not) in the 21st Century.* New York: Penguin, 2008.

Allen, W. He calls it a fact: State's quake expert believes in psychic phenomena. *St. Louis Post-Dispatch,* October 21, 1990.

Bernstein, S. How risky are older concrete buildings? *Los Angeles Times,* October 11, 2005.

Best, J. *More Damned Lies and Statistics: How Numbers Confuse Public Issues.* Berkeley: University of California Press, 2004.

Bolt, B. *Earthquakes.* New York: W. H. Freeman, 2006.

Braile, L. et al. Tectonic development of the New Madrid rift complex, Mississippi Embayment, North America. *Tectonophysics* 131 (1986): 1–21.

Braile, L., W. Hinze and R. Keller. New Madrid seismicity, gravity anomalies, and interpreted ancient rift structures. *Seismological Research Letters* 68 (1997): 599–610.

Brown, D. A shot in the dark: Swine flu vaccine's lessons. *Washington Post,* May 27, 2002.

Brown, M. Letter to the editor. *Memphis Commercial Appeal,* June 27, 2005.

Brumbaugh, D. *Earthquakes: Science and Society.* Upper Saddle River: Prentice Hall, 2010.

Byrne, R. Earth quacks. *St. Louis Riverfront Times,* December 5, 1990.

Calais, E. and S. Stein. Time-variable deformation in the New Madrid seismic zone. *Science* 5920 (2009): 1442.

Calais, E. et al. Tectonic strain in plate interiors? *Nature* 438 (2005): doi:10.1038/nature04428.

Calais, E. et al. Triggering of New Madrid seismicity by post-Pleistocene erosion. *Nature* 466 (2010): 608–611.

California Heath Care Foundation. *Seismic Safety: Will California's Hospitals Be Ready for the Next Big Quake?* Issue Brief, January 2007.

Camelbeeck, T. et al. Relevance of active faulting and seismicity studies to assess long-term earthquake activity in Northwest Europe. In *Continental Intraplate Earthquakes: Science, Hazard, and Policy Issues,* eds. S. Stein and S. Mazzotti. Geological Society of America Special Paper 425 (2007): 193–224.

Charlier, T. Quake-proofing costly, difficult. *Memphis Commercial Appeal*, May 25, 2003.

Coburn, A. and R. Spence. *Earthquake Protection*. New York: Wiley, 2002.

Cox, A. *Plate Tectonics and Geomagnetic Reversals*. San Francisco: W. H. Freeman, 1973.

Cox, R. et al. Paleoseismology of the southeastern Reelfoot rift in western Tennessee and implications for intraplate fault zone evolution. *Tectonics* 25 (2006): doi:10.1029/2005TC001829.

DeMets, C. et al. A revised estimate of Pacific-North America motion and implications for western North America plate boundary zone tectonics. *Geophysical Research Letters* 14 (1987): 911–914.

DeMets, C. et al. Current plate motions. *Geophysical Journal International* 101 (1990): 425–478.

DeMets, C. et al. Effect of recent revisions to the geomagnetic reversal time scale on estimates of current plate motion. *Geophysical Research Letters* 21 (1994): 2191–2194.

Dutton, D. It's always the end of the world as we know it. *New York Times*, January 1, 2010.

Elnashai, A. et al. *Impact of Earthquakes on the Central USA*. Mid-America Earthquake Center Report, August 2, 2008.

Ervin, C. and L. McGinnis. Reelfoot rift: Reactivated precursor to the Mississippi Embayment. *Geological Society of America Bulletin* 86 (1975): 1287–1295.

Farley, J. *Earthquake Fears, Predictions, and Preparations in Mid-America*. Carbondale: Southern Illinois University Press, 1998.

Feder, R. NU profs fault ch. 2 for earthquake alert. *Chicago Sun-Times*, November 28, 1990.

Federal Emergency Management Agency. *Designing for Earthquakes*. Publication 454, 2006.

Federal Emergency Management Agency. *HAZUS99 Estimated Annualized Earthquake Losses for the United States*. Publication 366, 2000.

Feynman, R. *The Pleasure of Finding Things Out: The Best Short Works of Richard P. Feynman*. New York: Basic Books, 2000.

Feynman, R. *What Do You Care What Other People Think? Further Adventures of Curious Character*. New York: Norton, 1988.

Finn, R. Rumblings grow about Parkfield in wake of earthquake prediction. *Nature* 359 (1992): 761.

Fischman, J. Falling into the gap. *Discover*, October 1992, 58–63.

Forte, A. et al. Descent of the ancient Farallon slab drives localized mantle flow below the New Madrid seismic zone. *Geophysical Research Letters* 34 (2007): doi:10.1029/2006GL027895.

Frankel, A. How can seismic hazard in the New Madrid seismic zone be similar to that in California? *Seismological Research Letters* 375 (2004): 575–586.

Frankel, A. et al. *Documentation for the 2002 Update of the National Seismic Hazard Maps Documentation*. U.S. Geological Survey Open-File Report 02-420, 2002.

Frankel, A. et al. *National Seismic Hazard Maps Documentation*. U.S. Geological Survey Open-File Report 96-532, 1996.

Fuller, M. *The New Madrid Earthquake*. U.S. Geological Survey Bulletin 494, 1912.

Geller, R. et al. Earthquakes cannot be predicted. *Science* 275 (1997): 1616–1617.

Gere, J. and H. Shah. *Terra Non Firma: Understanding and Preparing for Earthquakes*. New York: W. H. Freeman, 1984.

Geschwind, C.-H. *California Earthquakes: Science, Risk, and the Politics of Hazard Mitigation*. Baltimore: Johns Hopkins University Press, 2001.

Glassner, B. *The Culture of Fear: Why Americans Are Afraid of the Wrong Things*. New York: Basic Books, 2000.

Gordon, R. and S. Stein. Global tectonics and space geodesy. *Science* 256 (1992): 333–342.

Hager, B. et al. *Earthquake Research at Parkfield, California, for 1993 and Beyond, Report of the NEPEC [National Earthquake Prediction Evaluation Council] Working Group*. U.S. Geological Survey Circular 1116, 1994.

Hahn, R. and R. Litan. *An Analysis of the Second Government Draft Report on the Costs and Benefits of Federal Regulations*. Washington, D.C.: AEI-Brookings Joint Center, 1998.

Hamilton, R. and A. Johnston. *Tecumseh's Prophesy: Preparing for the Next New Madrid Earthquake*. U.S. Geological Survey Circular 1066, 1991.

Hanks, T. and C. Cornell. Probabilistic seismic hazard analysis: A beginner's guide. In *Fifth Symposium on Current Issues Related to Nuclear Power Plant Structures, Equipment, and Piping*, 1994.

Hebden, J. and S. Stein. Time-dependent seismic hazard maps for the New Madrid seismic zone and Charleston, South Carolina, areas. *Seismological Research Letters* 80 (2009): 10–20.

Henrion, M. and B. Fischhoff. Assessing uncertainty in physical constants. *American Journal of Physics* 54 (1986): 791–798.

Hernon, P. and W. Allen. Why many believed quake prediction. *St. Louis Post-Dispatch*, December 9, 1990.

Holbrook, J. et al. Stratigraphic evidence for millennial-scale temporal clustering of earthquakes on a continental-interior fault: Holocene Mississippi River floodplain deposits, New Madrid seismic zone, USA. *Tectonophysics* 420 (2006): 431–454.

Hough, S. *Earthshaking Science: What We Know (and Don't Know) about Earthquakes*. Princeton: Princeton University Press, 2002.

Hough, S. *Richter's Scale: Measure of an Earthquake, Measure of a Man*. Princeton: Princeton University Press, 2007.

Hough, S. Scientific overview and historical context of the 1811–1812 New Madrid earthquake sequence. *Annals of Geophysics* 47 (2004): 523–537.

Hough, S. and R. Bilham. *After the Earth Quakes: Elastic Rebound on an Urban Planet*. Oxford: Oxford University Press, 2006.

Hough, S. et al. On the modified Mercalli intensities and magnitudes of the 1811–1812 New Madrid earthquakes. *Journal of Geophysical Research* 105 (2000): 839–864.

Johnston, A. Seismic moment assessment of earthquakes in stable continental regions. *Geophysical Journal International* 126 (1996): 314–344.

Junger, S. *The Perfect Storm: A True Story of Men Against the Sea*. New York: Norton, 1997.

Kearey, P. and M. Brooks. *An Introduction to Geophysical Exploration*. Oxford: Wiley-Blackwell, 2002.

Kenner, S. and P. Segall. A mechanical model for intraplate earthquakes: Application to the New Madrid seismic zone. *Science* 289 (2000): 2329–2332.

Kerr, R. Earthquake forecast endorsed. *Science* 228 (1985): 311.

Kerr, R. Earthquake—or earthquack? *Science* 250 (1990): 511.

Kerr, R. From eastern quakes to a warming's icy clues. *Science* 283 (1999): 28–29.

Kerr, R. Parkfield earthquake looks to be on schedule. *Science* 231 (1986): 116.

Kerr, R. Parkfield keeps secrets after a long-awaited quake. *Science* 306 (2004): 206–207.

Kerr, R. Parkfield quakes skip a beat. *Science* 259 (1993): 1120–1122.

Kerr, R. Seismologists issue a no-win earthquake warning. *Science* 258 (1992): 742–743.

Kerr, R. Stalking the next Parkfield earthquake. *Science* 223 (1984): 36–38.

Kerr, R. The lessons of Dr. Browning. *Science* 253 (1991): 622–623.

Lawson, A. et al. *The California Earthquake of April 18, 1906: Report of the State Earthquake Investigation Commission*. Washington, D.C.: Carnegie Institution, 1908.

Levy, M. and M. Salvadori. *Why Buildings Fall Down*. New York: Norton, 1992.

Li, Q., M. Liu and S. Stein. Spatiotemporal complexity of continental intraplate seismicity: Insights from geodynamic modeling and implications for seismic hazard estimation. *Bulletin of the Seismological Society of America* 99 (2009): 52–60.

Liu, L. and M. Zoback. Lithospheric strength and intraplate seismicity in the New Madrid seismic zone. *Tectonics* 16 (1997): 585–595.

Liu, L., M. Zoback and P. Segall. Rapid intraplate strain accumulation in the New Madrid seismic zone. *Science* 257 (1992): 1666–1669.

Lovett, R. Quake analysis rewrites history books. *Nature* online (April 29, 2010) doi:10.1038/news.2010.212.

Lyell, C. *Principles of Geology*. London: Penguin Books, 1997.

Marshak, S. *Earth: Portrait of a Planet.* New York: Norton, 2007.

Marshak, S. and T. Paulsen. Midcontinent U.S. fault and fold zones: A legacy of Proterozoic intracratonic extensional tectonism? *Geology* 24 (1996): 151–154.

Mazzotti, S. Geodynamic models for earthquake studies in eastern North America. In *Continental Intraplate Earthquakes: Science, Hazard, and Policy Issues*, eds. S. Stein and S. Mazzotti. Geological Society of America Special Paper 425 (2007): 17–33.

Mazzotti, S. et al. GPS crustal strain, postglacial rebound, and seismic hazard in eastern North America: The Saint Lawrence valley example. *Journal of Geophysical Research* 110 (2005): doi: 10.1029/2004JB003590.

McKenna, J., S. Stein and C. Stein. Is the New Madrid seismic zone hotter and weaker than its surroundings? In *Continental Intraplate Earthquakes: Science, Hazard, and Policy Issues,* eds. S. Stein and S. Mazzotti. Geological Society of America Special Paper 425 (2007): 167–175.

Medawar, P. *Advice to a Young Scientist.* New York: Basic Books, 1979.

Meert, J. and T. Torsvik. The making and unmaking of a supercontinent: Rodinia revisited. *Tectonophysics* 375 (2003): 261–288.

Mitchell, B. Memorial to Otto Nuttli. *Bulletin of the Seismological Society* of *America* 78 (1988) 1387–1389.

Mitchell, M. *Complexity: A Guided Tour.* Oxford: Oxford University Press, 2009.

Mussett, A. and M. Khan. *Looking into the Earth: An Introduction to Geological Geophysics.* Cambridge: Cambridge University Press, 2000.

Nevada Bureau of Mines and Geology. *Living with Earthquakes in Nevada: A Nevadan's Guide to Preparing for, Surviving, and Recovering from an Earthquake.* Special Publication 27, 2000.

New York Times. Aquarium wins FEMA pay for fishing trips. November 21, 2007.

Newman, A. et al. New results justify discussion of alternative models. *Eos, Transactions of the American Geophysical Union* 80 (1999): 197.

Newman, A. et al. Slow deformation and lower seismic hazard at the New Madrid Seismic Zone. *Science* 284 (1999): 619–621.

Newman, A. et al. Uncertainties in seismic hazard maps for the New Madrid Seismic Zone. *Seismological Research Letters* 72 (2001): 653–667.

Nuttli, O. W. The Mississippi Valley earthquakes of 1811 and 1812: Intensities, ground motion, and magnitudes. *Bulletin of the Seismological Society of America* 63 (1973): 227–248.

Obermeier, S. Liquefaction evidence for strong earthquakes of Holocene and latest Pleistocene ages in the states of Indiana and Illinois, USA. *Engineering Geology* 50 (1998): 227–254.

O'Connor, C. Human chromosome number. *Nature Education* 1 (2008): 1–4.

Oreskes, N. Why predict? Historical perspectives on prediction in earth science. In *Prediction: Science, Decision Making and the Future of Nature*, eds. D. Sarewitz, R. Pielke, Jr. and R. Byerly, Jr. Washington D.C.: Island Press, 2000.

Parsons, T. Lasting earthquake legacy. *Nature* 462 (2009): 42–43.

Penick, J. *The New Madrid Earthquakes*. Columbia: University of Missouri Press, 1981.

Pilkey, O. and L. Pilkey-Jarvis. *Useless Arithmetic: Why Environmental Scientists Can't Predict the Future*. New York: Columbia University Press, 2006.

Robbins, W. In quake zone, a forecast set off tremors. *New York Times*, December 1, 1990.

Robbins, W. Midwest quake is predicted: Talk is real. *New York Times*, August 15, 1990.

Sagan, C. *The Demon-Haunted World: Science as a Candle in the Dark*. New York: Ballantine Books, 1996.

Sarewitz, D., R. Pielke, Jr. and R. Byerly, Jr., eds. *Prediction: Science, Decision Making, and the Future of Nature*. Washington D.C.: Island Press, 2000.

Savage, J. The Parkfield prediction fallacy. *Bulletin of the Seismological Society of America* 83 (1993): 1–6.

Schweig, E., J. Gomberg and J. Hendley. The Mississippi Valley—"whole lotta shakin' goin' on." U.S. Geological Survey Fact Sheet-168-95, 1995. http://quake.usgs.gov/prepare/factsheets/NewMadrid/.

Searer, G., S. Freeman, and T. Paret. Does it make sense from engineering and scientific perspectives to design for a 2475-year earthquake? In *Continental Intraplate Earthquakes: Science, Hazard, and Policy Issues*, eds. S. Stein and S. Mazzotti. Geological Society of America Special Paper 425 (2007): 353–361.

Sella, G. et al. Observations of glacial isostatic adjustment in stable North America with GPS. *Geophysical Research Letters* 34 (2006): doi: 10.1029/2006GL02708.

Sexton, J. et al. Seismic reflection profiling studies of a buried Precambrian rift beneath the Wabash Valley fault zone. *Geophysics* 51 (1986): 640–660.

Siegel, M. *False Alarm: The Truth about the Epidemic of Fear*. Hoboken: Wiley, 2005.

Sieh, K. and S. LeVay. *The Earth in Turmoil: Earthquakes, Volcanoes, and Their Impact on Humankind*. New York: W. H. Freeman, 1998.

Sieh, K., M. Stuiver and D. Brillinger. A more precise chronology of earthquakes produced by the San Andreas fault in southern California. *Journal of Geophysical Research* 94 (1989): 603–624.

Singer, P. Why we must ration health care. *New York Times Magazine*, July 19, 2009, 38–43.

Smalley, R. et al. Space geodetic evidence for rapid strain rates in the New Madrid seismic zone of central USA. *Nature* 435 (2005): 1088–1090.

Snay, R. Horizontal deformation in New York and Connecticut: Examining contradictory results from the geodetic evidence. *Journal of Geophysical Research* 91 (1986): 12,695-12,702.

Snay, R., J. Ni and H. Neugebauer. Geodetically derived strain across the northern New Madrid Seismic Zone. In *Investigations of the New Madrid Seismic Zone, 1538*-F, eds. K. Shedlock and A. Johnston. U.S. Geological Survey Professional Paper, 1994.

St. Louis Post-Dispatch. Experts debate Missouri fault's seismic activity. December 15, 2005.

St. Louis Post-Dispatch. It can happen here. July 25, 1990.

Stein, R. Earthquake conversations. *Scientific American*, January 2003, 72–79.

Stein, R. The role of stress transfer in earthquake occurrence. *Nature* 402 (1999): 605–609.

Stein, S. Comment on Frankel, A. How can seismic hazard in the New Madrid seismic zone be similar to that in California? *Seismological Research Letters* 76 (2005): 364–365.

Stein, S. No free lunch. *Seismological Research Letters* 75 (2004): 555–556.

Stein, S. Understanding earthquake hazard maps. *Earth*, January 2009, 24–31.

Stein, S. and M. Liu. Long aftershock sequences within continents and implications for earthquake hazard assessment. *Nature* 462 (2009): 87–89.

Stein, S. and S. Mazzotti, eds. *Continental Intraplate Earthquakes: Science, Hazard, and Policy Issues.* Geological Society of America Special Paper 425 (2007).

Stein, S. and A. Newman. Characteristic and uncharacteristic earthquakes as possible artifacts: Applications to the New Madrid and Wabash seismic zones. *Seismological Research Letters* 75 (2004): 170–184.

Stein, S. and E. Okal. Speed and size of the Sumatra earthquake. *Nature* 434 (2005): 581–582.

Stein, S. and J. Tomasello. When safety costs too much. *New York Times*, January 1, 2004.

Stein, S. and M. Wysession. *Introduction to Seismology, Earthquakes, and Earth Structure.* Oxford: Blackwell, 2003.

Stein, S. et al. Midcontinent earthquakes as a complex system. *Seismological Research Letters* 80 (2009): 551–553.

Stein, S. et al. Passive margin earthquakes, stresses, and rheology. In *Earthquakes at North-Atlantic Passive Margins*, eds. S. Gregerson and P. Basham, 231–260. Dordecht, Netherlands: Kluwer, 1989.

Stein, S., A. Newman, and J. Tomasello. Should Memphis build for California's earthquakes? *Eos, Transactions of the American Geophysical Union* 84 (2003): 177, 184–185.

Sunstein, C. *Risk and Reason.* Cambridge: Cambridge University Press, 2002.

Swafford, L. and S. Stein. Limitations of the short earthquake record for seismicity and seismic hazard studies. In *Continental Intraplate Earthquakes: Science, Hazard, and Policy Issues*, eds. S. Stein and S. Mazzotti. Geological Society of America Special Paper 425 (2007): 49–58.

Tuttle, M. The use of liquefaction features in paleoseismology: Lessons learned in the New Madrid seismic zone, central U.S. *Journal of Seismology* 5 (2001): 361–380.

Tuttle, M., H. Al-Shukri and H. Mahdi. Very large earthquakes centered southwest of the New Madrid seismic zone 5,000–7,000 years ago. *Seismological Research Letters* 77 (2006): 755–770.

Twain, M. *Life on the Mississippi*, 1883.

Van Arsdale, R. *Adventures Through Deep Time: The Central Mississippi River Valley and Its Earthquakes*. Geological Society of America Special Paper 455 (2009).

Wang, Z., B. Shi and J. Kiefer. Comment on Frankel, A. How can seismic hazard in the New Madrid seismic zone be similar to that in California? *Seismological Research Letters* 376 (2005) 466–471.

Weinstein, L. and J. Adam. *Guesstimation: Solving the World's Problems on the Back of a Cocktail Napkin*. Princeton: Princeton University Press, 2008.

Winchester, S. *A Crack in the Edge of the World: America and the Great California Earthquake of 1906*. New York: Harper Collins, 2005.

ACKNOWLEDGMENTS

This book grows from research done with coworkers, fruitful discussions with them and other researchers studying these or related problems, and knowledge from the broad geological community. In that spirit, I thank many people for making this book possible. All should feel free to share credit for parts of the book with which they agree and to disclaim parts with which they disagree.

The first is my wife, Carol. We've been a two-geologist family for 28 years, with all the fun and complications that poses. Her willingness to support this book and encourage me when progress slowed is even more impressive given that she went through it all for my previous book.

Next are my colleagues at Northwestern, especially Donna Jurdy, Emile Okal, Brad Sageman, and Suzan Van der Lee. Contrary to the incorrect image that outstanding research scientists aren't interested in promoting public understanding of science, everyone was fully supportive of the time and effort that went into the book.

I thank my coauthors on the research papers discussed here, including Don Argus, Eric Calais, Sierd Cloetingh, Mike Craymer, Chuck DeMets, Tim Dixon, Roy Dokka, Joe Engeln, Andy Freed, Richard Gordon, James Hebden, Tom James, Qingsong Li, Mian Liu, Ailin Mao, Glenn Mattiolli, Stephane Mazzotti, Jason McKenna, Andres Mendez, Andrew Newman, Emile Okal, Paul Rydelek, Giovanni Sella, Carol Stein, John Schneider, Norm Sleep, Laura Swafford, Joe Tomasello, Roy van Arsdale, John Weber, and Rinus Wortel. I also thank the many students and staff from different institutions who helped with the GPS field program.

I also have benefited from discussions with Amotz Agnon, Rick Aster, Gail Atkinson, Thierry Camelbeeck, Dan Clark, Nick Clark, Jim Cobb, Bill Dickinson, Alex Forte, Anke Friedrich, Tom Hanks, Bob Hermann, Sue Hough, Ken Hudnut, Alan Kafka, Steve Kirby, Cinna Lomnitz, Mike Lynch, Steve Marshak, Brian Mitchell, Orrin Pilkey, Jim Savage, Gary Searer, Bob Smalley, Bob Smith, Ross Stein, Zhenming Wang, Steve Wesnousky, Michael Wysession, and many others.

Finally, I thank Patrick Fitzgerald and Bridget Flannery-McCoy at Columbia University Press for supporting the project and getting the manuscript into a book.

INDEX

Note: Figures and tables are indicated by *f* or *t*, respectively, following the page number.